Sam McAughtry was born in 1921, in the loyalist Tiger's Bay area of north Belfast, one of a family of ten. He left school at fourteen and worked at a variety of unskilled jobs, including a spell as a riveter in Shorts aircraft factory. In the war he joined the RAF ground staff and, after study, became an officer navigator. After the war he was, among other things, a builder's labourer in London and a car-body worker in the Ford plant at Dagenham. In 1980, after having been a part-time newspaper correspondent and feature-writer for many years, he left a Civil Service job to write full time. He is now well known for his weekly column in the *Irish Times* and for his television and radio broadcasts on both sides of the Irish border.

McAUGHTRY'S WAR

SAM McAUGHTRY

THE
BLACKSTAFF
PRESS
BELFAST AND DOVER, NEW HAMPSHIRE

The author is grateful for permission to quote from 'Easter 1916' by W. B. Yeats. Lines reprinted by permission of M. B. Yeats and Macmillan London, Ltd.

First published in 1985
by The Blackstaff Press
3 Galway Park, Dundonald, Belfast BT16 0AN, Northern Ireland
and
51 Washington Street, Dover, New Hampshire 03820 USA
with the assistance of
The Arts Council of Northern Ireland

Typeset by The Brown Fox
Printed in Northern Ireland
by The Universities Press Limited

British Library Cataloguing in Publication Data
McAughtry, Sam
McAughtry's War.
I. Title
828'.91407 PR6063.A119/
Library of Congress Cataloging-in-Publication Data
McAughtry, Sam
McAughtry's war
1. McAughtry, Sam. 2. World War, 1939-45—Personal narratives, Irish. 3. Great Britain. Royal Air Force—Biography. I. Title.
D811.M395 1985 940.54'81'415 85—22950

ISBN 0 85640 352 0 (hardback)
0 85640 339 3 (paperback)

And so our scene must to the battle fly;
Where, O for pity! we shall much disgrace
With four or five most vile and ragged foils,
Right ill-dispos'd in brawl ridiculous,
The name of Agincourt. Yet sit and see;
Minding true things by what their mockeries be.

The Life of King Henry the Fifth
Act 4 Scene 1

Foreword

In order to sharpen my recollection of the war years I revisited some of the places mentioned in this book before sitting down to write it. Forty-five years to the day after I left home to enlist in the Royal Air Force Volunteer Reserve, I took the Larne-Stranraer ferry and journeyed again through Scotland to Padgate, in Lancashire.

There is no reception centre in Padgate any more, only bare and sour ground where once recruits were broken in on the barrack square. There is just the faint outline of the paths where I once walked, carrying stiff, ill-fitting clothes. Twined about my fingers were identity discs that were said by the sergeant to be fireproof. Thank God they weren't put to the test.

I revisited St Athan and there is still an RAF station there. I stood in the very cookhouse that served 500 airmen in 1940. It was bombed by the Germans in the summer of that year, with me inside. In tiny St Athan village I sat again in the same Four Bells pub, in the same seat where I sat in 1940 and lowered my first pint.

Then I drove to Cranfield in Bedfordshire and was delighted to find that the Cranfield Institute of Technology had retained many of the old wartime buildings on the campus. The guardroom is still there, intact, and the clerical staff sat most respectfully silent as I stood in the very cell in which I was once locked up for arson about in working hours. It was nice to stand there, with my eyes closed, and pray silently that the RAF police who looked after me then had had a nasty war.

At the RAF College Cranwell I was taken to the College dining hall and lunched by the officers, under oil paintings of Prince Charles, Lord Trenchard and Group Captain Peter Townsend. In the corridor was a faded photograph of Group Captain Douglas Bader, the legless war ace, who was a cadet at Cranwell in the Thirties. I was taken to see again the spot on the airfield perimeter where once I saw a crashed Percival Proctor aircraft, with, just visible, the carbonised heads of a staff pilot and one of my cadet comrades. And it was good of the modern day RAF police to salute me as I left Cranwell. I always felt superior to them in my young day on the station, and it was nice of the present generation to accept this assessment.

In Babbacombe, Devon, on a glorious day, I traced the hotel where, in 1942 when it was requisitioned by the Air Force, I crammed and sweated on the first stage of a navigator's course. It was the first time that I had ever studied for anything in my life. The window corner where I sat, puzzling the mystery of the parallelogram of forces, is still there, just the same, today. So is the spot in the garden where, in another heat wave, I put on the heavy flying clothes and posed, under false pretences, for a snapshot. I wondered then whether I would ever wear them in earnest. Well, I did, and as I stood a lifetime later in the same place, I could remember well the uncertainty of a time when I was a roughneck in the company of college boys.

On the Scottish island of Islay it is not at all difficult to find exciting links with the war. The airfield that the RAF built is a civil airport today. The wreckage of no fewer than three wartime Beaufighters can still be seen nearby. Their crews, over forty years dead, were friends and comrades of mine.

Islay is a peaceful, lovely place now, and during my return visit its people were most kind and hospitable to me. It seemed unbelievable that I once hated the island and its black, November storms, rushing in without warning from the Atlantic. There is a historian on the island who has records of every aircraft crash that took place there during the war. One of mine earns half a line, and it's the sort of immortality I could have been doing without.

In 1944 I flew with a Beaufighter squadron from the old Western Desert bases, carrying out attacks on shipping and harbour installations in the Aegean islands to the north. I never forgot their beauty. For this book it was essential that I should revisit at least one of the islands, and there was never any doubt as to which one it would be . . .

My taxi breasted a high hill, then, for the downward run, its nose dropped, as the Beaufighter's had done on the attacking dive, and there before me, its houses still white, still bright in the sun, was the little town of Samos. Its harbour was empty, and quiet. By God, it was anything but quiet the last time I'd seen it. The German gunners were trying to blow us out of the sky, and we were just about to pulverise a ship full of mines. But that was long ago, and when I found Manolis Panomaris, who had been the one to radio from Samos for the RAF to come, the two of us sat and talked old soldier talk, about times when we were young and thought that we'd never grow old.

In Sliema, Malta, I found the phone number of my old transport sergeant, Salvo. We'd been through many good and bad times in Greece when that country was recovering from grievous damage following German occupation. I was in a state of high excitement. Salvo and I would meet and talk over old times . . .

He'd been forty-three in 1944, and I'd been twenty-three. There's not much difference between forty-three and twenty-three. We were alive and full of energy, then. But add forty-one years, and the difference can't be measured, it's so great. Salvo didn't even remember me. He didn't remember Greece, except for the fact that he'd been there. He was an old man, and I'd disturbed his rest. So I apologised, and said that I understood. Then I hung up, remembering him as he once was.

After my sentimental journey it wasn't hard to write this book. Although there's war in it, it isn't just a war story. It's more a story about me, and I don't think it's very complimentary, but at least I hope it's honest.

I have altered names here and there, because some of the people in the book must still be alive. In other cases I haven't mentioned names at all. The names of my friends who died are real, though.

The final draft was done at the Tyrone Guthrie Centre, Annaghmakerrig, County Monaghan. No more appropriate place to commemorate forty years of peace can be imagined.

The Arts Council of Northern Ireland has given generous assistance. This is much appreciated. I am also grateful to that most civilising influence, the *Irish Times* of Dublin, for making the journey to my wartime landmarks possible.

1

Off to the war

When I was growing up there were old soldiers and young soldiers all around. I was born in 1921, so the litter and leavings of the 1914-18 war lay everywhere, and Cosgrave Street had its fair share of them. I remember seeing Army web equipment nailed to the side posts of front doors to make swings for toddlers, and up above the doors, behind half-moons of glass, steel helmets hung, with green plants trailing from them.

The men from the trenches in Flanders were put down more trenches in the Belfast streets, and instead of being paid in cash they were given vouchers to redeem in grocers' shops. The shopkeepers weren't allowed to include cigarettes in the order, but most of them turned a blind eye. Ex-soldiers were powerful smokers.

Well into the Twenties the work was scarce and the men crowded the street corners. Sometimes they would scrape together a few coppers and start up a pitch-and-toss school. I earned many a penny keeping lookout for the cops but it was necessary to look out for more than the cops: my vantage point was the backyard wall of a neighbour's house and, since the womenfolk sometimes peed in the yard, the sight of a wee boy watching the proceedings was liable to invite a bucket of water. The more touchy women were given soldiers' nicknames by the local men: one was known as Big Bertha, after a famous howitzer used by the Germans, and another bucket-happy lady was called Moaning Minnie. This was the name given by the infantry to the German *Minenwerfer* bomb hurled across the front-line trenches.

We had shell-shocked men who talked to phantoms. And near our house there was a man called Gerry who worked all day and coughed all night, for he'd been gassed at Neuve Chappelle in 1915. Gerry's wife had run away from him and their two daughters. The girls, fag-smokers just into their teens, ran the streets wild and had to be chased home, the neighbours playing grouse-beaters, every night, while Gerry stood wild-eyed and whispering at his front door.

Once he had them in the house Gerry coughed nothing but abuse at the two girls, and walls were so thin that every word could be heard. It made Mother wince, the things that were said, but I knew, and my brothers, that the accusations hurstling past Gerry's lips as the pair were beaten upstairs were absolutely true. Their knickers were down more than they were up, and maybe it wasn't all that surprising, in such a loveless home. Poor Gerry was really a hospital case, but he had to work to feed the girls, and it killed him before he was forty.

War talk was still going on into the early Thirties. Mother often told us about her youngest brother, Billy, lost in the Battle of the Somme in 1916. The word came through that he was missing in action. Only days later a local man, known as Gunner Gordon because he'd been a machine gunner in the Connaught Rangers with Billy, came home to Belfast on leave. Mother saw him in the street.

'The wire from the Army only said our Billy was missing,' she said. 'What does that mean?'

'Well he's bound to be missing, isn't he?' Gunner Gordon said, brightly. 'He was blew to smithereens, wasn't he?'

Mother collapsed in the street. When she came round she never forgave Gunner Gordon, the poor, blunt soldier, still talking the language of the trenches. When Gunner's name was mentioned in our house, Mother mouthed words like 'pig' and 'porcupine', one of her strongest oaths.

Sammy, another brother of Mother's, was cut down at the Somme as well, wounded by machine-gun bullets. He was a small man, and I often saw the scars on his shoulder and chest as he washed at his scullery jawbox. I wondered at the toughness of him, able to pull through after such a stitching.

Over and above these reminders of war, young men all around us were being sucked into the peacetime Army, in the absence of other employment. It was the Royal Ulster Rifles that took delivery of our fellows in the main. After an absence of three months or so they would come back stiff-necked, head-back automatons, on embarkation leave for India or Palestine. The tour of duty overseas was seven years. Nobody in the district thought anything of a young lad going away from his parents and family for seven years. That was the Army, and there was no more to be said.

For myself as a boy there was the Church Lads' Brigade, run, like other church youth organisations of the time, on strict militaristic lines. We were even drilled by sergeants of the King's Royal Rifles, the regiment in quarters in Victoria Barracks at the time. Each year we held our drill competition at – of all places – the Cripples' Institute on the Donegall Road. The Belfast Battalion would carry out the most complicated marching and wheeling manoeuvres, forming fours and reverting again to twos, opening order for inspection, marching past in Guards order. I was no good at the drill caper myself, but all my days I've admired those who were.

There were no pacifists in Cosgrave Street. We thought highly of the Crown forces. When World War Two was more than a year off the men of my district joined the Territorial Army in hundreds. The annual bounty of eight pounds or so helped, of course, but the men would have joined anyway. The Army was a part of our background, and the 1914-18 war casualties to be seen on every side did nothing to dim our enthusiasm.

In September 1939, when Neville Chamberlain announced on the radio that Great Britain was at war with Germany, my mother slid back in her chair and fainted. Since our sister wasn't there at the time, I and two of my brothers ran next door to Mrs McClean for help. Kindly Mrs McClean brought Mother round. 'I know why you're upset, Mrs McAughtry,' she said, 'so go ahead and cry.'

And Mother cried because her brother Billy had been lost at the Somme, and because her man was a merchant seaman, sailing with my brother Mart, and because my brother Tommy had told her that he was going to join the North Irish Horse, and because I had said that I was going away in the RAF as soon as I could manage it. Mother had plenty to cry about on Sunday morning 3 September 1939.

I was working as a riveter in Shorts aircraft factory in Belfast. It was a reserved occupation, but I was speedily taken off the reserved roll when I was caught hammering the nipple of a rivet-gun air-line to loosen it, making it easier to unscrew. This practice had been banned by written order a short time before. Summarily sacked, I was at once assured by the shop steward that my fellow riveters were all for going out on strike in sympathy with my cause. 'We don't think you should get the beetle just for hammering a nipple,' the shop steward said.

'Don't bother, I'm joining the Air Force,' I told him. 'I never want to see another nipple as long as I live.'

Coming from a sex-starved, eighteen-year-old virgin, it must stand as one of the daftest exit lines ever uttered.

We were standing in the LMS railway station on York Street on the afternoon of Monday 19 February 1940, waiting for the Larne-Stranraer boat train. I was carrying a cardboard suitcase containing a pair of drawers, a pair of socks, two hankies, and enough fried egg and bacon sandwiches to feed a bomber squadron, this being Mother's idea of something to keep me going between real meals. 'Tell the man that I can send you soda bread over if they have none in England', she said on parting. Uncle Alec had an even bigger parcel of egg and bacon sandwiches, and the funny thing was he had a bad stomach and couldn't eat fried food, but in our connection it would have been unthinkable to get on the boat without taking a 'fry' for the journey.

It was a coincidence that Uncle Alec was making the crossing with me. He was going over to join the Pioneer Corps. My brother Jack, who was leaving us down to the train, was puzzled about this. 'After all that you went through in France, what the hell are you enlisting for?' Jack asked. It was a very good question. Uncle Alec had fought with the Connaught Rangers, the Royal Navy and the Royal Flying Corps in the first war. He'd been a very poor soldier, doing time in a military prison in Carrickfergus Castle, deserting the Army to join the Navy immediately on his release, and then deserting the Navy to join the Flying Corps. Seemingly he had used his real name and address in all three enlistments, and they'd let him away with it because he had a row of gunshot wounds across his ribs and stomach from his days in the trenches, and anyway the authorities recognised a character when they saw one.

During his twenty-one days' detention in Carrickfergus Castle Uncle Alec had been under the impression that his Army pay of one shilling a day would be saved for him. Instead, he was paid only a penny a day. On the day set aside for his release, when he was offered one shilling and ninepence instead of twenty-one shillings, he'd flattened a sergeant, knocked a first lieutenant onto the floor, and had been hauled back to the cells for a further three months. 'A penny a day, is it?' he'd yelled, as they took him away. 'Well how's that for a pennyworth?' No wonder Jack was surprised to see him going back to the Army.

'Look, mate,' Uncle Alec replied, 'with my Army experience they'll make me a sergeant in the Pioneer Corps in no time. All this regiment does is dig latrines, so there's no brainwork involved. I wouldn't be surprised but what, when this war's over, I'll be a regimental sergeant major. I think I'll make a lifetime career out of this.'

As usual, he turned out to be wrong to the power of ten. What the Army did was to put him to digging fortifications in France. When the Germans invaded Belgium and France Uncle Alec was tearing along in lorries and trains and horses and carts only yards in front of the leading German tanks. He only just made it to Dunkirk. When he got back to Belfast he was so out of breath that the Army gave him his discharge, and this time the Navy and Air Force wouldn't take him, so for the next five years until the end of the war he spent his time writing to his sons and nephews in the armed forces, giving them advice on how to dodge the column. But we weren't to know all that on 19 February 1940.

I had never in my life slept away from home. I had never travelled more than twenty miles from our house. I knew none of the RAF party travelling to England with me, and I hadn't the slightest idea of what lay ahead. My trade was to be flight rigger, and that had been chosen for me because my record showed that I'd worked in an aircraft factory for a year. The Air Force had yet to learn that I was a terrible riveter. I was dreadfully seasick on the crossing to Scotland. Uncle Alec had fallen in with some other old soldiers and was up at the bar singing songs from the first war. Every now and again he'd come back to where I was sitting, alone. 'Bring it up,' he'd advise, 'you'll feel a different man.' I'd already done so, and didn't think much of the different man I was feeling, but Uncle Alec was a kindly chap, genuinely in sympathy with the young, and his concern did help.

We were put on a train in Stranraer, near midnight. Some Inniskilling Fusiliers going back to France set up the old Army game crown-and-anchor, and the gambling voltage brought me back to life as nothing else could have done. Early on the morning of 20 February I fell asleep, with the pleasurable feel of win-money in my pocket. Somewhere near the border with England Uncle Alec woke me. 'Good luck, Sam,' he said. 'And always remember, volunteer for nothing.' Then he was gone, and I was on my own.

In an upstairs room in the attestation centre, RAF Station Padgate, Lancashire, I fell asleep, exhausted after the boat and train journey. I was awakened by the sound of a voice calling 'Paddy, Paddy.' I opened my eyes slowly, wishing that Paddy, whoever he was, would answer his name and give the rest of us peace. Looking around, I saw the rest of the draft of civilian recruits smiling at me, waiting for me to acknowledge the sergeant's summons.

It was the first time I'd ever been called such a name. Well, naturally. Round about where I came from, Paddies tended to be as rare as crucifixes at the Presbyterian General Assembly. Oddly, coming from a district which had provided so many men for the King's colours, nobody had ever mentioned the fact that in the armed forces all Irishmen were Paddies, all the way up to and including officers of Air rank.

'Yes?' I struggled to my feet, content to be Paddy. For the next six years I would be known by no other name. When I was with other Irishmen I used their proper Christian names, but it was also common for the Irish to address each other as Paddy.

I liked it. To be Irish is to be special and interesting. As far as I could see, other Northern Irish Protestants reacted in the same way. I was to see Antrim Presbyterians come out of themselves, begin to show a swash here and a buckle there within days of signing on, because the British expected it of them and because it offered a lovely release of the inhibitions, considering the times we were in.

A corporal who lived in the little room at the end of the hut where I lived was a Monaghan man, a regular who had served for several years and had just returned from a tour in Iraq. He went out of his way to befriend me from the outset. At a time when I felt not only lost and lonely, but awkward and stupid, in the midst of what I took to be sophisticated Britishers, Tom Brody nicely and quietly eased the hard days for me.

A drill corporal had made fun of me in front of the crowd, and I'd whaled him for it, earning myself fourteen days confined to camp. Tom showed me how to do up my web equipment for inspection at the guardroom, calmed me down. 'It's not worth it, Paddy,' he said. 'Wait till you get the bastards somewhere on their own, then give them a quiet warning. That way you get peace and quiet.'

'There's some'll go for the Irish if they can,' he said to me once. 'Don't make it easy for them.' In the canteen, Brody had more advice for me. 'Go and buy a writing pad and write to your mother,' he said. 'Get into the habit of it. When you get your twenty-four bob a fortnight, send her about six. It's always something, and it'll please her that you're thinking of her.'

Peter Kane, the world flyweight champion, was a gym instructor at Padgate. Brody introduced me to him. 'Paddy's from Belfast,' he said to Kane, 'where Jimmy Warnock, the southpaw, lives.'

Kane smiled. 'A lovely left hand he has too,' he said.

Tom had picked a time when the other recruits and the drill corporal whom I'd run across were watching. I was grateful. It made me feel a good deal less of a slum Sammy, black at the ankles, shirt and no drawers.

I can remember Padgate for very little, other than the warm discovery of my Irishness. It was like being a member of the Masonic Order. I used to run down and corner the owner of every Irish accent, North or South, that stood out from the British. I would find out where in Ireland he came from and I would pull out the fags or buy beer and talk about home. I kept doing that right through the whole war. A Londoner with the name of Patrick Murphy drew the Red Hand of Ulster on the front of my steel helmet, and underneath it, in Celtic script, he wrote 'Uladh'.

Drawn by the emblem of Ulster, people from home came up to me on station platforms and in transit camps halfway round the world, and introduced themselves to me as Irish. Later I had that helmet stolen from me, along with my gas mask. It happened in a Church Army canteen in Swindon. As soon as I discovered the loss I stole a gas mask and steel helmet from a soldier of the Ox. and Bucks. Regiment. Weeks later a mock gas raid was sprung on us at the airfield. As soon as the rattles sounded I tore open the gas-mask case. Inside it, instead of the respirator I expected, was a complete shaving outfit, rolled up in a towel. A passing warrant officer went purple as the soap and shaving brush were exposed to the light of day. He made one angry bound towards me, but relented when he heard my Belfast tongue.

It turned out he came from near the North Wall in Dublin. He was most interested to learn that my old dad's ship tied up there, at the Palgrave Murphy berth. Knew the place well, did the warrant officer. Described it to me in detail. Forgot all about the shaving kit in the gas-mask case.

On St Patrick's Day, which fell the day before I was due to leave Padgate as a fully uniformed, injected, inoculated, hair-trimmed, shoulder-straightened airman, I was given my shamrock by Corporal Brody, and a pin with which to fasten it onto my forage cap. Feeling quite superior, I went into Warrington town with a recruit from the Falls Road called Pat Johnston. We'd been told all about the bromide in our tea for the last month, but it was either a myth or it was too weak for my system. I was feeling razor-sharp in the sex line. We hadn't even properly left the bus stop when we were picked up by two crackers.

'Is that shamrock? Are you Irish?'

Mine was small and cuddly, that's all I knew. 'Sure and be Jasus and it's ourselves that's the lucky ones to be after meeting with two such gorgeous colleens,' says I, offering my arm. It was taken with alacrity. We took them to a café, since not one of the four of us took drink. My senses were swimming, once we all sat down. I was getting the footsie treatment from Cuddles. Pat, a serious-minded sort, was embarrassed at the way I was shoving on the stage-Irish, but it seemed to be working. 'Have you any snapshots, Terence?' my girl asked, using my chosen *nom de guerre*, Terence O'Toole. It so happened that I had. I fished out a blurred picture of myself taken one Easter Tuesday just before I'd marched off with the Juvenile Orange Order to the field at Bangor, County Down. In the snap I was wearing sash and white gloves and gauntlets, for I was the lodge secretary.

Cuddles was over the moon. 'You're an Orangeman?' Modestly, I admitted it.

'Could you write your name and address on the back?' Most certainly I could. This relationship was going at a mile a minute. Where I came from, the exchange of snapshots was a serious business. I signed it, 'With love – Terence O'Toole'. But Cuddles hadn't a snap to give me in return. Nor had the ash-blonde with Pat, even though he'd given her one of his and entered his name on the reverse side too.

Tea and buns over, I began to fidget. 'What about going to some park or the pictures or something?' The footsie had me steamed up. But the girls weren't keen. Indeed, alarmingly, they looked like baling out. And that's exactly what they did, and nothing we could do would stop them.

As they left, Cuddles turned to Blondie: 'That's an Irish, a Canadian, a Welsh and a Scotsman I have. Wasn't it lucky, meeting them?'

On the train going to St Athan in Glamorgan next day, to learn to be flight riggers, Johnston and I agreed that it had been worth it, if only for the knee-nudging. It was the first time it had happened to us.

At St Athan we were each put to work shaping a rough piece of mild steel into a cube, using only a file. The cube had to be near perfect. When this was finished, we had to shape two plates into squares, cut two smaller square holes in each, and set the cube into both plates so that the fit was tight at all points. I was damn nearly kicked off the course when the instructor saw my early work, but luckily, after I started in to drinking, I met a guy in a bar in Barry nearby, and solved the problem. This chap was a corporal fitter on the permanent staff and he took a great shine to me in the bar on account of the fact that I bribed him with drink. 'I'll get you one of the cube and plate sets handed in by an earlier course,' he said, and he did too, rescuing me from disgrace. I used my influence to get one for Alec McCleod too, a man from the Isle of Skye, who was my closest mate. Alec was as handless as I was, and that's saying something, but I couldn't see him kicked out for the sake of a bit of cheating.

We were unlikely pals. He was pure Highland middle class. He'd interrupted an arts degree course at Edinburgh University to join up as a rigger. He introduced me to literature, and to the noble art of pint-drinking. In the Four Bells inn just outside the main gate he would talk to me about university life, and I couldn't get enough of it. I knew nothing about the literature of my own country. In the long, lovely spring of 1940 Alec McCleod tried to put that right as best he could. Every time I see in print Yeats's elegy for the poets and rebels of 1916, I can hear Alec's Skye sibilants:

> Hearts with one purpose alone
> Through summer and winter seem
> Enchanted to a stone
> To trouble the living stream.

His hair was red-brown, he stood six feet and had a solid, hardy, islander's independence. His accent made the most commonplace remark seem significant to me. In the Cardiff bars we had what was to me unbelievable success with the girls. I was in the tail of the comet right enough, but that was where I wanted to be.

One day, when yellow-painted Anson reconnaissance aircraft were taking off and landing, we were both waiting to go up on our first flip – ten minutes in the cockpit of a Miles Magister. A few yards from us an officer stood, one of the operational Anson crews. He wore working dress uniform, with the trousers tucked into leather flying boots. Around his neck, held by the intercom lead, was a flying helmet. He was someone from another world. This was one of the gods who would fly the planes that I would one day service. I saw that Alec was watching the officer closely.

'That's what I'm going to be one day,' he said.

'What does he do?'

'He's the observer. He navigates the aircraft, and drops the bombs.'

The thing about it was, Alec looked the part. He was every inch the part. He was university material, just arsing about as a rigger, arsing about, killing time with bohunks like me till the day came when others of his class would beckon, and then he would go and take his rightful place, in a shiny barathea uniform. When his aircraft was taking off to bomb the Germans, I'd be saluting at the dispersal, in my blue dungarees, with my screwdriver in my hand. Suddenly the flip in the Miles Magister was downgraded. It seemed more like the charabanc runs that we were given as kids, by the church, as a charity.

Up in the air, over the scattered convoy of tiny ships returning to the Bristol Channel from the Dunkirk beaches, I spoke to the pilot: 'Did you go to university?'

'Christ, no,' he laughed. 'I left school at sixteen.'

'Does an observer need a university degree?'

'Not at all. Their entrance standards are similar to pilots'.'

It made me feel a little better. There wasn't all that much difference between leaving school at fourteen, like me, and at sixteen, like an observer. It wasn't an impossible gap to close. It seemed ridiculous at the time, but I found myself wondering whether I could handle the observer's job.

Meanwhile, the rigger's minimum pass mark was forty per cent. I got forty-one and McCleod achieved forty-two. This meant an immediate pay rise from two shillings a day to three and sixpence. We went out to Barry to celebrate. I woke up next morning with a black eye. In the next bed, McCleod had a sprained wrist. We both had skinned knuckles. The Army had blamed us for the defeat in France. We had given up our rooms to the soldiers from Dunkirk and we'd bought them beer in Barry. Then, when they were drunk, they'd turned on us. Seemingly there were no fighters over the beaches.

'What happened to all my money?' I wanted to know.

'You gave it to the Salvation Army,' Alec said. 'You fell for the girl selling the *War Cry*.'

Next day we set off for No. 14 Service Flying Training School at Cranfield, Bedfordshire. We were to be let loose on Airspeed Oxford trainers as qualified flight riggers, aircraftmen second class. Between the pair of us we could hardly have fitted the top on a fountain pen. If we'd been sent to work on Spitfires, the Battle of Britain could have been lost.

2

Prison and other alarms

Cranfield was a well-appointed peacetime RAF station situated in the rural area of Bedfordshire. Only seven miles from Bedford, it lay exactly halfway between Oxford and Cambridge. With other new arrivals, Alec McCleod and I were driven to camp in an open fifteen-hundredweight truck, through beautiful English hamlets and villages with gorgeous names: Apsley Guise, Wharley End, Ampthill and Woburn Sands. I was most impressed, though, by the size of the English arable fields. Back home our farms were made up of tiny jigsaw pieces, but in the vicinity of the airfield the growing crops seemed to extend as far as the very horizon: hundreds of acres of produce and nothing else, no hedges, no loanings, no rutty lanes, and most certainly no crazy, hanging Irish gates. And every once in a while, by the roadside, the English pubs, still with their coach-house names, like the Leathern Bottel. In time I was to become such a regular at the Leathern Bottel that I knew all about the family troubles of the landlord, and I could pull strings for bitter beer when it became scarce, but on the way to Cranfield that first time in August 1940, I was in much better mood to see and enjoy the surroundings, since no training course and no exams awaited me, and I was enchanted by everything that I saw.

Apart from the women, this was why I had joined up. To see the world. Everything was brand new to me. My letters home – and I wrote home every other day for the six years I was away – were full of descriptions of my new surroundings. The quiet people of Bedfordshire would have been surprised at the eagerness with which I passed on to my Belfast relations my impressions of this peaceful part of England.

Alec McCleod and I were given accommodation in one of the brick-built barrack blocks, and we had no sooner settled down than in walked a little guy with one of the cockiest bantam walks I'd seen since Saturday night on York Street in Belfast. His name was Jack

Burgess; he was indeed from Belfast; he was a flight mechanic and, like Alec and me, his rank couldn't have been lower.

Right from the start he surprised me by saying that he was going to be a pilot one day.

'Do you not need education for that?' I asked.

'No,' he replied, 'all you need is brains.'

Fair enough, says I to myself. The world and its ways were becoming clearer.

In next to no time the three of us found ourselves on airfield guard. Ground-staff tradesmen were expected to stand their turn on guard at that time, in the face of the threat of German invasion. We were given five rounds of .303 rifle ammunition, told there was no more where it had come from, and sent out to patrol the perimeter track, in twos.

On the very first shift, Burgess managed to get away from his partner and join Alec and me behind the civilian canteen, where the McAlpine labourers ate. I stood on Alec's shoulders and gained entry. Afterwards we gobbled the roast beef dripping sandwiches which were all I could find in the place, and cursed our luck. A Glasgow man called Jake Thomson had told us that he had discovered a huge steak and kidney pie the previous night and had eaten the lot himself. When we got back we told Jake that we'd found a dozen custard and a dozen coconut buns, and we pretended to be sick with the richness of them. Later, in a bar in Bedford, Thomson told us that he'd lied about the steak and kidney pudding. He'd only found roast beef dripping sandwiches, and Burgess, McCleod and I looked disgusted at his lies.

An Airspeed Oxford crashed in a field in the heart of the Bedfordshire countryside, and Burgess, McCleod and I volunteered to stand crash guard and prevent the local yokels from stripping the thing for souvenirs. We dumped our blankets in a nearby barn, stood guard until it was dark, then, against orders, the three of us headed for the nearest pub. The handful of farmers inside the ancient inn were amazed to discover that we had been the crew of the Oxford when it hit the deck.

'Talk about lucky to get away with it,' Burgess mopped his brow. 'If it hadn't been for Alec's experience as a test pilot with Handley

Page we'd have been goners.'

'We 'eard there were on'y one man in 'er, an' 'e were a dead sergeant,' some old swede-basher said.

'Every time we crash it's the same thing,' I said. 'The whole countryside's full of rumours.'

'Why dew yew 'ave tew stay 'ere, then?' the barmaid enquired. 'Won't they let yew goo back to th' airdrome, then?'

'Secret equipment,' Burgess explained. 'We're here to guard it until the scientist comes to see if it's damaged.'

We got back to the crash, free beer scooting out of us, and not one of us stood a minute's guard through the night. Watches were resumed at nine the next morning. Burgess fired twice at rabbits in the next field, and got seven days confined to camp for wasting ammunition when we got back. His excuse that our rations were insufficient was not accepted.

While Burgess was doing his punishment I developed a tremendous respect for him. During the long hours in the barrack room he proved to be a ruthless poker player, the very opposite of myself. Upon the arrival of a dull-witted Welshman who had just drawn six months' back pay for a good conduct stripe, Jack first lured the airman into a quiet little game of nap, for halfpenny stakes, then, taking all the time in the world over it, he enlarged the school by inviting Alec and myself to join. In a couple of hours, the game had become poker. At two o'clock in the morning, when our eyes were stinging with lack of sleep, the Welshman rose from the table. He hadn't a penny left of his good conduct pay. He was pathetically grateful for the five Woodbine that Burgess tossed his way. For each week that the poor bastard remained in our billet, Burgess took half of his wages off him at poker, before the Welshman even had a chance to go out and drink.

'It's doing him a kindness,' Jack said, 'the Welsh are like Red Indians. Firewater's the worst thing they can get.'

At Cranfield I met up with a red-haired Belfastman who lived near me. His name was Tom O'Connell, and he came from North Queen Street, about ten minutes' walk from my own district. Of course, in religious and cultural terms, since he was a working-class Catholic, he might as well have come from Heckmondwyke in

Lancashire. O'Connell and I met when we were paired on airfield guard right over the weekend. This meant that we stood two hours on and four off for two and a half days, spending twenty strucken hours walking the perimeter. In that time, despite the fact that we hadn't one complete page of Irish history held tight in our heads, we went into the National Question, dropped, drowning, in its centre, and came out the other side, having learned to support each other on the way.

It began with religion. Naturally.

'You have to admit it, mate, you worship graven images. I've seen them. Statues all over the place.'

'We don't worship the statues. They're there to help us to pray.'

'That's another thing, crossing yourself. You cross yourselves going past the candy apple. Is that so you'll know each other?'

'What's that? What's the candy apple?'

'The chapel.'

'Wait a minute, mate. I don't make fun of your church. Why should you make fun of mine?'

'You mightn't make fun of our church, but you think I'm a heathen. I was out with a wee girl from your street one time and she told me that: "Youse Prods are only heathens. Yez have no religion, so yez haven't." That's what she told me.'

'Well neither yez have. The Catholic faith is the only true faith.'

'If you say that again I'll hit you a thump on the frigging gub.'

'I'm only having a discussion. Not that you frigging worry me, mate. Nobody from across the Duncairn Gardens would worry me. But it's only a discussion, you know.'

'Billy Grant says that youse ones has far too many kids.'

'Who's Billy Grant?'

'He's something at Stormont. He declared the painting of King Billy open, on Lilliput Street gable wall.'

'Well we were told that Protestants use surgical rubber goods. That's a sin in our church.'

'Well my oul' fella never used any surgical rubber goods. There was ten kids in our family.'

'Well, when this Billy Grant said that Catholics had too many kids, did you not tell him there was ten in your family? I suppose you just wanted to hear us being criticised.'

'If there's one thing I'll never understand it's all this blessing yourselves. You'd think you owned the cross or something. Other people recognise the cross too, you know.'

'Well why don't you recognise it right, by blessing yourselves the same as us? It was a cross that Our Lord was killed on, you know. It wasn't a circle. If it hadda been a circle we would bless ourselves with our hands going round and round. What's the matter with that? And so should you.'

'If they'd shot him, would all you Fenians stick out your fingers and go "Bang, Bang", when you're going past the candy apple?'

At this point, halfway through the first two-hour watch, O'Connell and I were on the side of the airfield furthest away from the hangars and buildings. We were at the mouth of the runway. Night-flying activities had just begun, and goose-neck paraffin flares were spluttering into life along the runway's edges. But we couldn't have cared less. O'Connell had had enough, and it was quite all right with me if he wanted to show it. He threw his Lee Enfield down on the grass. Mine went down on top of it, and the pair of us got stuck in.

It's a sickening feeling to find that an opponent has fast hands. O'Connell was one lovely mover. He had my nose streaming blood and one eye shut within twenty seconds. Trying to get close nearly cost me a mouthful of teeth, but close I got, in time, and in went knees, head, gouging fingers, and down went the boot on top of his toes, with the shins taking their quota of punishment *en route*.

'Fight clean', he gasped.

'D'ye think I'm frigging mental,' I came back. The lad was a boxer, way above my class.

I had him down and was setting about choking the very life out of him, when suddenly a fierce, blinding glare shone down on us, as from an avenging god.

Struggling on the perimeter track at the runway entrance, we rolled apart and, paralysed with fear, turned our eyes to heaven. We were caught in a light that was brighter than any light we had ever known. Simultaneously there came a sound that was like the end of the world and a huge, menacing shape hurtled down from the night sky. Terrified, O'Connell and I hugged each other for protection.

With a snarling, spitting rush an Airspeed Oxford tore just above our heads. As its wheels screeched on the tarmac surface the hot slipstream covered us in dust and fine stone particles. Blinded, in total darkness, we lay for a long minute, catching our breath. Then we stood up, dusted ourselves off, picked up the rifles, and resumed our joint patrol.

'Christ, you didn't half hurt my shins there. You're a rough bastard, McAughtry.'

'Bloody near thing, that kite landing there. Christ, O'Connell, wouldn't it have looked funny, a letter from the Air Ministry to your oul' woman: "Dear Madam, The Air Ministry regrets to inform you that your son was killed fighting a Prod, on 21 October last. An Oxford lit on the pair of them, and if it's any consolation to you, your son was winning the fight when he was squashed as flat as a pancake."'

'"Dear Mrs McAughtry, Your son Sam was shot down by one of his own aircraft on 21 October. However, you can tell the rest of the Prods in your neighbourhood that he took a Fenian with him."'

'Is that eye closed right over?'

'It is. When we go over to the cookhouse, I'll ask for a lump of meat to hold on it.'

'My trousers are torn. I'm in for it from the corporal of the guard.'

'Me too. But I'm not worried. Up the corporal from here to Carrickfergus.'

'You haven't a fag on you, have you?'

'Aye, here. Here's a light.'

The Tom O'Connell story turned out to be a brief one, as far as I was concerned. He was unskilled, and the station warrant officer took a dislike to him. To a general duties airman, this was calamitous. Poor Tom was given the most demeaning work going, including latrines duty. He came to me for advice. I suggested that he should apply for trade training, or even aircrew, but the SWO's approval was needed for both. Eventually O'Connell solved the problem in the Irish way. He went over the wall next time he took leave, deserting to Dublin.

Before heading off he called up at my home in Hillman Street and told Mother that he was stationed with me, and was on his way back to Cranfield after leave. God love her, she packed up a big parcel of fried egg and bacon and soda farl sandwiches and gave it to O'Connell to give to me. She also sent a half a crown for me. O'Connell took the lot and blew off to Dublin with it. Before leaving, he dropped me a line:

'I needed every deuce I could get to keep me in Dublin. I knew your mother would give me dough for you. Some day I'll send you the half dollar. And the soda bread wasn't wasted. See you after the war.'

A year later they caught him. There was no work in Dublin. He came North, got a job as a truck driver, and found himself delivering material to Victoria Army Barracks next door to where he lived. The soldier at the gate demanded his identity card and of course he had none. They gave him fifty-six days in the military prison known as the Glasshouse, at Aldershot, and when he came out he was so totally brainwashed that he became an RAF policeman.

Early in 1946 we met in a Belfast bar. O'Connell bought me two pints for the half crown, and a glass of whiskey for the fried egg sandwiches.

On 13 November 1940, the SS *Gloucester City* docked in St John's, Newfoundland, carrying ninety-two survivors of a naval action fought in the middle of the North Atlantic. The German pocket battleship *Admiral Scheer* had intercepted a convoy of thirty-eight merchant ships, designated HX84, under the sole protection of an elderly passenger liner, the *Jervis Bay*. The escort was pounded by the warship's eleven-inch guns until she sank, a blazing hulk, with the loss of her captain and 189 of her Navy crew. Then the German captain set about the destruction of the convoy. But in the end he only managed to sink five merchantmen. A combination of bad timing, excessive caution on the enemy's part and high valour on the Royal Navy's side, leading to the award of a posthumous VC to the escort's captain, Fogarty Fegen, restricted the *Admiral Scheer*'s bag considerably.

The SS *Gloucester City* had, amongst its survivors, half the crew of the *Kenbane Head*, my brother Mart's ship. Dad and he had signed on with separate ships as soon as war broke out. When the rescue vessel reached St John's, the names of the survivors were telegraphed home to Ireland. Mart's name was not there. He was lost.

I didn't know this when news of the convoy battle came over the BBC radio. With Alec McCleod and Jack Burgess I sat in the barrack in Cranfield and talked about the sea fight. The others respected my views on the engagement. As the grandson, son and brother of merchant seamen I knew something about the subject. But I didn't know that the brother to whom I was so very close had been killed by a shell fragment when a lifeboat had been hit before it could be lowered to the sea from the deck of the *Kenbane Head*. When I received the news from home one afternoon I ran out through the camp to the very airfield perimeter, and threw myself down on the grass and cried my eyes out.

Later, in the evening, Alec took me out to Bedford town. We went into the Carrington Arms. After an hour or so I got up off my seat and went for a walk, alone, through the town. By the river I found the church of St Paul, and I went inside, to the Trinity Chapel. There, alone, I sat, with nothing happening in my mind.

I didn't feel God's presence. No voice assured me that Mart was happier where he was. Nothing at all in the way of comfort came my way. After I don't know how long, I rose and went back to McCleod in the pub. On the way, it occurred to me that Mart had always thought highly of me. He had always seemed to think that I had brains. Mother had once told me that he showed my letters around the crew in the fo'c's'le of the *Kenbane Head*.

Here I was, an aircraftman second class, the lowest rank in the Air Force. One way that would help to make things up to his memory would be for me to gain promotion. It mightn't at the same time be a bad idea to set about punishing the Germans for what had happened. The two objectives came together in the single aim of becoming an observer. But then something happened to hold that particular business up for a while.

In December 1941 Burgess and I came near going to the Glasshouse ourselves. Burgess set fire to the dope store at one end of the hangar, and it damn nearly sent the whole hangar up in smoke, with the aeroplanes in it.

I was the rigger on duty crew, which meant that I was responsible for looking after visiting aircraft, and in fact on the day in question I had come down to the flight where Burgess and I normally worked in order to tell him that I had just sent off an aircraft flown by Amy Johnson, the celebrated record-breaking woman pilot. Full of this, I ran Jack to earth in the dope store. This was where the paint used on the aircraft was kept. All around the floor were large tins of it, and under the window were two pails of dirty petrol, used to clean the paint off the hands after use. Jack was acting the goat with a Scotsman called Thompson, dipping his fingers in the petrol and flicking it over him. The place was reeking of petrol when I went in. Thompson went out for some reason and, seeing me, Burgess instinctively pulled the cigarettes out of his pocket, preparatory to catching up with my gossip. I turned to go outside for the drag but, to my horror, Jack struck the match inside and the place went up like a torch.

Burgess lost his eyebrows and some of his hair. His tunic jacket flamed and he ripped it off as we ran away from the fire. About ten yards off we turned. The dope-store windows burst like bombs as the pails of petrol caught. Flames were shooting skywards. 'Christ!' Jack shouted, 'There's an Oxford the other side of the fire in the hangar.'

'Frig it,' says I, pulling him. My instincts were telling me to get offside. But Burgess was already on his way back. As the station siren screamed, he grabbed an extinguisher and appeared to walk straight into the heart of the fire, into the dope store. Cursing him for a stupid bastard, I went in after him, lifting an extinguisher myself.

He had gone through to the hangar side of the store and was directing the jet from the extinguisher to the wall on that side. 'Don't open the door,' he shouted to the fitters inside the hangar, 'keep it shut.' Together we sprayed the interior of the dope store, but it was only possible to stay for seconds, before the heat drove us out. Soon the station fire tender had arrived, and the blaze was brought under control and confined to the dope store.

Seemingly a panicking fitter inside the hangar had been about to open the door and risk the fire reaching an Airspeed Oxford trainer parked only feet away, when Jack's voice had stopped him. But never mind all that, I was off and away to the duty crew room under the control tower as soon as the fire crew turned up. I didn't stop to say cheerio, just took off in obedience to my gutter senses.

At lunchtime I turned up in the cookhouse, all innocent, drew my grub, and sat beside some mates from the flight. 'Your pal's locked up,' they said. Burgess had been lifted by the RAF police within minutes.

I thought about things as I munched the sawdusty sausages. Burgess would never, in a million years, incriminate me in the thing. I could go on about my business. Nobody had seen me enter the dope store. In any case, the entire blame was Jack Burgess's. The bloody idiot had struck the match that started the fire. What the hell was I feeling so guilty about?

I rose, went to my barrack room, sat on the bed that was next to those of McCleod and Burgess, stared out of the window. In a short time Alec McCleod came in. 'Have you heard about Burgess?' He was out of breath.

I nodded. 'Have you seen any sign of my name being mixed up in the fire?' I asked.

He shook his head. 'Only that you helped to put it out. Apparently it was Thompson who put the finger on Jack. But Thompson himself was lifted. They're both in the station guardroom.'

With a heavy heart I packed my small kit, towel, toothbrush, razor, soap. Then I took my cigarettes out, took a razor blade, sliced about six fags into tiny, half-inch parts and hid these about my person, together with the heads of a dozen matches. McCleod watched sympathetically. 'You'll not be in long,' he said. I sighed, shook hands with him, and went to report to the guardroom.

'I was in the dope store when it went up,' I said to the thin-lipped sergeant in charge.

'Good, Paddy,' he said. 'Your mate doesn't look the same without you.' I was shown to a large cell.

Burgess and Thompson were sitting glumly at opposite ends.

Burgess jumped to his feet. 'What the hell are you doing here?'

I waited till the sergeant had gone. 'I'm here to fight your case,' I said.

We were locked up for five days altogether. There was a preliminary enquiry. At first things looked rough. One squadron leader, a special investigation officer, remarked on the fact that Burgess and I were Irish. 'Are your hearts in this war?' he wanted to know. 'It wouldn't be the first time that the Irish acted as fifth columnists against the British.'

Jack and I were standing in line together, facing the group captain, a kindly-looking man; another squadron leader; and the man who had questioned us. Thompson was in another room. I waited for Jack. His reply was nervous. 'We're not against the British,' was all he could say.

'If crime is proven, you'll go to the Glasshouse for at least a year,' the officer said. I felt myself growing very angry. The squadron leader saw it in my face. 'Did something I said annoy you?' He was staring at me with narrowed eyes.

I swallowed. I wasn't so long in the RAF that I could take on a senior policeman and stay cool.

'I think you've got a bloody brass neck,' I said suddenly, feeling the head going away from me altogether. The squadron leader half rose, but the group captain put a hand on his sleeve.

'Only last month my brother was lost in the merchant service,' I said. 'Right now my dad's in the frigging Atlantic Ocean. He was torpedoed in the last one. There's no torpedoes anywhere around you that I can see, nor ever will be either. There's no conscription in Ireland, but we're here, and there's not all that many English volunteers alongside Burgess and me, let me tell you. This man wants to be a pilot, that's how much of a fifth columnist he is. It seems to me that the more Irishmen you can get the quicker you'll win the bloody war.'

I was totally out of breath, and terrified as well at my temerity. In the silence I felt sure that my heart's pounding was making a noise like a Lambeg drum.

'You did set fire to Air Ministry property,' the group captain said.

'And Burgess nearly killed himself putting it out again,' says I.

'Take a look at him, sir.'

Your man's eyebrows were gone entirely. He wore bandages on both hands. Purple anti-burn liquid stained his forehead. He had the instinctive good sense to say nothing. He simply stood to attention as he was inspected by the officers.

After being sent to wait outside, we were recalled and told that it had been decided not to send the papers for court martial action.

'Instead you will be confined to barracks for a further nine days, making fourteen days' punishment in all,' the station commander said. It was nice to see that Thompson, the informer, got the same.

When I resumed duty I got down at once to the business of remustering for aircrew training. I soon got permission to attend lectures with the pupil pilots on the camp. Instead of hanging around the flights playing pitch and toss while the Oxfords were flying, I went up in the air, looking over the pupils' shoulders as they learned to read maps and charts. Soon I was able to make sense of latitude and longitude, courses and tracks, compasses and gyros.

I had a great deal of help in this direction from a girl called Martha, daughter of a woman with whom I had been billeted in Woburn Sands village. Martha's mother was a widow. She was hard of hearing. As a consequence of this, she slept soundly the night Martha seduced me and I lost my virginity at long last.

When I'd lived in the barrack room the talk every night amongst the other guys had always been of their conquests. I was the only one in the room, seemingly, still to connect. The funny thing was that I kept quiet about the relationship between Martha and myself once it progressed to its nice, natural limit. I would study for hours each night in the back bedroom of the semi-d. in Woburn Sands, then, when the mother went to bed, Martha would slip in, and my studies would take another turn. The regime seemed to do me good. The lovemaking sealed the other studies nicely in my mind.

3

The reluctant volunteer

The main problem with me was my reputation. The man with the biggest say in the matter of my transfer to flying training was Tommy Roe, the engineering officer, a man who had gone through the ranks, and was nobody's mug. Unfortunately, he regarded me as the ringleader of our trio. The applications by Burgess for a pilot's course and by McCleod for an observer's course were greeted sympathetically. They were merely put on probation for several months as a precondition for appearing before an aircrew selection board, but in my case Tommy Roe stipulated that I should prove my capacity for obeying orders by earning two jumps in promotion as a flight rigger. I could go forward to the selection board when I had reached the rank of leading aircraftman.

The problem could hardly have been bigger. I was easily the worst rigger in the Air Force. I left things inside aeroplanes when I'd finished working on them. Whole vital parts, like ailerons, could have been left off during my pre-flight inspection and I'd never have noticed. All around, as the RAF expanded, riggers were gaining promotion, but I was never considered. I still tended to hammer things off if they seemed too tight to unscrew. So some way, other than a facility with spanners and wrenches, had to be found to get me over the engineering officer's stiff condition.

I took the sergeant out for a drink. I liked him and he liked me. He was a Sunderland man, with a city face.

'The question of rapid promotion arises, Geordie,' I said.

'It arises, Paddy,' he replied, 'and then it subsides again, to the ground. Flat.'

'I want to talk to you about the civilian shortage of high-tension radio batteries,' I said. 'I want to talk about this, after I've bought you whisky.'

'Speak,' said Geordie, three minutes later.

'They're throwing these HT batteries out in the radio section when they're only half spent,' I said. 'It's something to do with safety tolerance.'

'What's the going price in the pubs for a half-spent HT battery?'

'There's this Derryman living in Bedford who'll be here in a wee minute to tell us.'

Geordie took a swig from his whisky glass, studying me. 'Does it mean so much to you to go flying, Paddy?'

'It does.'

'I've never been able to understand such a thing. Good luck to you anyway.'

It was a month before the next promotion board was held on the station. My name was put up on orders. 'Even if you don't get it,' Tommy Roe said, 'at least you've been called. That's a record in itself.'

'Oh, I'll get it all right,' says I. The HT battery business was going great guns.

The Sunderland sergeant took the board: 'D'ye see this, fastened to the mainplane?'

'I do.'

'What's its function?'

'It activates the barometer for the airspeed indicator.'

'Congratulations. You're an aircraftman first class.'

A month after that: 'D'ye see this here? What's it do?'

'It activates the barometer for the airspeed indicator.'

'Congratulations. You're a leading aircraftman.'

The funny thing is, I actually improved at the job when I was promoted.

In April 1941 Belfast had its first heavy air raid. I was home on leave at the time. I brought some souvenirs with me. They were tail fins from incendiary bombs which had been given to me by a fitter who lived in a part of London where the Luftwaffe were bombing every night. For myself, the only enemy aircraft I'd seen were lone Ju 88 or Dornier raiders, whose pilots must have been hopelessly lost to have drifted over Bedfordshire. There was hardly an operational airfield in the whole of the county. At night-time we could see the London blitzes: blue, bright pinpoints of flak twinkling in the far sky; searchlights hoping to grope Heinkels, making senseless angles with beams that were yellow to us, blue to the aircrews. Ripples that started in Hammersmith and Plaistow reached the soles of our boots in Bedfordshire and Buckinghamshire every night.

'That's the sort of stuff that falls on us all the time, over there,' I said, on the Easter Monday. The tail fins were passed round the family and the lads at the street corner, and there was much whistling as they were examined. 'I'll put them up on the mantelpiece,' my young brother said. 'They'll make great souvenirs.'

I was leaving a woman called Susie Lavery home from the York Road fleapit on Easter Tuesday night, and hoping to get her into the parlour of her house, when there was this droning in the sky.

'I'm afraid,' Susie Lavery said, at her door.

'Oh, so am I,' I said. 'Come on and we'll go into your parlour and keep each other's spirits up.'

But there was nothing doing. I was summarily dismissed, and before I had even reached home the bark of Bofors guns had started, the sound hitting the walls of the narrow streets and bouncing, and filling everywhere with a wild, howling aftersound.

The Luftwaffe Dorniers, Heinkels and Ju 88s homed onto their targets by way of crossed radio beams called *Knickebein*. By this time we had moved to Hillman Street, and that night I swear the two signals met exactly over our house. When the bombing started all the payloads seemed to be heading for our chimney. It was something fierce. The house shook to its foundations and down past them. The air in our kitchen was full of plaster and cement dust and soot. Mother and my sister were in the coal hole. Since I was a fully-fledged member of the armed forces I thought it would look bad to hide in the coal hole, so I was cowering on the sofa, trying to sound blasé. 'It's all right,' I shouted, 'its bark's worse than its bite. In the morning you'll be surprised how little damage there is.'

When morning light broke, the truth of it was that Hillman Street and the whole area in which it lay looked like Passchendaele, but too many other things had happened for my forecast to bounce back at me. For one thing, my old dad was home, his ship having been torpedoed in the North Atlantic. At fifty-nine years of age, he'd had to scramble from his bunk up onto a wildly tilting deck, jump into the lifeboat, and sit with the other survivors in the open boat, praying that another ship would pick them up. This happened in due course. He was taken to Iceland and given a suit that was feet too big for him. Mother said that he looked like somebody from the music hall when he turned up at the front door a couple of weeks later. Not the least put out by all this experience, Dad landed up at the house right in the middle of the air raid with his protective arm around the local publican, a man of his own age, who was jerking and trembling and praying all at the same time.

'I've brought Jimmy round,' Dad announced, removing half-bottles of John Jameson from his pockets. 'Jimmy supplied this drink in return for me looking after him,' he explained. Thereafter, for the duration of the blitz, my old seadog dad sat on a chair singing Irish songs to Jimmy who was down on the floor under the table, with his rosary beads a blur, and praying like a tape recorder speeded up.

The John Jameson notwithstanding, it was a fact that my old man didn't seem to know the meaning of fear. Later in the war a young engineer shipmate of his told me that at the North African landings in 1943 Dad stood on deck watching with enormous academic interest as Ju 88 bombers tried to hit their convoy. 'He was wondering if that would have been the sort you would be flying against, Sam,' the engineer said, shaking his head, 'and the thing was, the ship that we were on was stuffed to the deck plates with ammunition, and the bloody bombs were missing us by only yards.'

Meanwhile, ashamed of myself for doing nothing, I decided to put my tin hat on and make an effort. The scene outside was so appalling that I simply put my head down and stopped looking around at it. I made my way around to Spamount Street nearby, where there appeared to be a great deal of rescue activity. By this time I had worked out that the bombs swishing down and failing to explode were incendiaries. Little pinpoints of light were appearing everywhere, as the first ignition began.

'Hey, mister!'

A man had rushed out of one of the houses halfway down the street. He was beckoning to me. There was no one else near.

'Yes?' I was hoping that I looked like Lew Ayres, the film star, in my tin hat down over one eye. My demeanour was confident, stern, battlehardened. 'Be careful of that shrapnel,' I warned.

The man grasped my arm. 'There's a fire in the front room,' he shouted. In our district the 'front room' was the front bedroom. He pointed up to it. Through the glass could be seen the beginnings of a fire.

'Right', I said, 'just calm down.'

He was in his late fifties, small, and delicate looking. He was in jacket and trousers, his shoes were unlaced and he wore no shirt.

'Whatever you do, don't get in my way,' I said, briskly, as we ran to the house. 'These incendiary bombs need careful handling. Where's your shovel?' The man ran through to the yard and handed me a small fireside shovel. I took it from him, weighed it expertly, and pointed to the street. 'I want no amateurs near me when I'm handling this,' I said. 'Go out there and wait.'

He pulled himself away. 'I'm fucking sure I'm not,' he snorted. Then he went in underneath the stairs. 'Tell me when you've fixed it.' By this time the smoke was curling throughout the house.

I ran upstairs, threw open the front-room door. A phosphorous bomb had entered through the roof and ceiling, burnt its way through a bed, and was spluttering, in the heart of a white blaze, on the floor.

I took a deep breath. It nearly killed me. My lungs seemed full to the top with ready-mixed cement. The fumes of the mixture before me were near-lethal. I'd been lectured on this a half-dozen times and had clean forgotten. Somehow I lifted one leg after another, convinced that I was a dead man, and made my way to the landing. I couldn't get more than a tiny trickle of air into my lungs, and the effort to do even that was superhuman. My one thought was to get air. Finding myself beside a tiny landing window I lifted my foot and smashed it through glass and woodwork. Then I put my head out into the wild night and tried to breathe. Gradually I recovered. In about five minutes I stood up again. The fire in the front room seemed no bigger. I made to go back in, holding my breath this time, but found to my surprise that I couldn't move my right leg.

In smashing the window I had ripped the back of my right ankle open to the bone. It was as though I had no foot at all. Somehow I hopped downstairs. The householder came out from below the stairs: 'Are you all right?'

The shovel was still in my hand. I gave it to him: 'What you do is you try to get that underneath the incendiary, and then you carry it downstairs and out onto the concrete road,' I said. I could feel the blood thumping out of my ankle.

'Did you not put my bloody fire out?' the man screamed.

I left him, hurling abuse after me, as I made the street. Whole sticks of bombs of the high explosive kind were falling again all around. I dropped to the ground and inched along, the way I'd seen the actors do in World War One films. I seemed to have the district to myself. Inching along proving too slow, I got on one foot, hung onto the wall, and tried to hop, but eventually it came to crawling, for the hopping made me twist the good ankle as I stepped on some rubble.

With bombs and landmines exploding all around and the shrapnel from anti-aircraft guns striking sparks from the streets, I crawled somehow round to the house again, where Mother stemmed the blood with a rough dressing until the raid died down. A policeman had meanwhile taken refuge in our house, refusing point blank to go out into the raid, in the face of my needling from the sofa. When the last aircraft had gone, the constable, shamefacedly, promised to alert the first-aid post about my injury, and he was as good as his word. An hour after the All Clear sirens a nurse arrived in a light ambulance, and I was taken to the Mater Hospital on the Crumlin Road first, then, when the doctors there found that the damage was relatively slight, I was taken to Victoria Barracks, Belfast Garrison HQ, and eventually driven from there to Campbell College in east Belfast, a public school recently converted into a military hospital.

When I was in the emergency reception area of the Mater Hospital the dead were being brought in and laid out, all around me. For the most part they weren't too distressing to see – blanket-covered bundles all in a row. There was too much bustle going on, too many doctors, drivers, nurses stepping over them. But the blanket had come down from the head of one that was near me: the face was turned away, and all I could see was a woman's hair. It was thick with mortar and brick dust but even so I could see that it was black, shot through with grey, going white at the crown of the head. It could have been Mother's hair.

That was when my head swam, and I dropped it between my knees and kept it there, until the nausea passed. Some Air Force doctor had passed this tip along, in a lecture at Cranfield. I was glad that I knew it, and had avoided fainting, being in full uniform at the time and very much aware of it.

I was three days in hospital before anybody came to visit me. The Army stitched my ankle and told me that I was extremely lucky that my Achilles tendon had escaped with only a nick, otherwise I'd have been a cripple for life. Lying in bed I told myself that it mightn't have been the worst thing in the world. 'Wounded in an air battle,' I might have been murmuring, to the girls back in Tiger's Bay.

Mother came to see me, out of breath, wearing her good coat with the fur collar, carrying a huge can filled with broth and insulated by a two-inch-thick application of the *Belfast Telegraph*, tied with wool. For some reason there was never any string in our house.

In the bed next to mine there was a cockney soldier, about forty years old, suffering from flat feet, and suffering is the right word. He'd had an operation for the condition, and the pain must have been fierce, for he'd moaned for three solid nights with it. Mother was sympathetic when I told her about his feet. 'I know, mister,' she said, 'I know what bad feet are like. Mine have been killing me for years. Did you ever try alum in hot water for it?'

The soldier and I polished off the broth. When she kissed me and left, the Londoner shook his head. 'My Gawd but it's lovely to meet folk like my own. So kind-hearted, like.'

I shrugged. 'Over here we would do it for a dog. This is Ireland you're in.'

'Give me the real working clawss,' the cockney said, staring at the door through which Mother had disappeared. 'People wiv nuffink. They're the salt of the bleeding earf.'

'What the frig do you think you're talking about?' I shouted. 'You might have nothing, back in cockneyland, but my family owns a bloody shipping line. Jesus Christ, but that's a good one,' I burst into shrill laughter. 'People with nothing. I live in a house with a frigging waiting room, you know.'

'A waiting room?' The soldier was baffled. 'Do you live in a bleeding railway station, or something?'

By the time he said it I had recovered my position. I always did get mixed up with dining rooms and living rooms and breakfast rooms and the like.

'Our house used to be a doctor's house,' I explained. 'We still call one of the living rooms the waiting room. OK?'

'I have been wounded in action,' I wrote to McCleod and Burgess. 'Come and visit me. I want to find out the truth. The doctors won't tell me.'

McCleod couldn't make it. He was away on detachment and couldn't be replaced, but three days later Jack Burgess turned up at my bedside. He'd told the Air Force he was going to some place in England for the weekend, and made it, somehow or other, to Belfast.

'How did you manage the boat and train fare?' I wanted to know.

'What you do is buy a ticket for a few coppers to some place nearby, a small station,' he explained. 'This destination is written by hand on the ticket. Then you take your bottle of Milton antiseptic, as used to sterilise baby's bottle, and apply it to the inked entry on the ticket. By some magical means this removes the name. It is simple then to enter "Belfast" on the ticket.' His face was full of concern, though. 'What's all this about your wound? Do you want me to see the doctor?'

I smiled bravely. 'I think it's too late, old chap,' I said. 'I've already lost the right leg, and they're taking the other one off after I've had my tea tonight. Tell McCleod I leave him my gumboots.' I finished weakly, 'You can have my socks.'

'I have worse news than that for you,' Burgess said. 'There was some woman called Susie Lavery at your house when I called there an hour ago on my way here. She had some notion of visiting you, but I told her that you were injured in the testicles and felt sensitive about it. The lassie is going out with me. Tonight I'll be showing her exactly where you were hurt.'

But I got my own back on him. When I got back on my feet I went to see Susie. 'Jack's crazy about you,' I said, and gave his full RAF address to her, as well as his home address. 'His mum wants to see him married,' I said. 'Why don't you go up to her house and get friendly.' It took months before he could stop the letters of adoration from Susie and the ones of protest from his mother.

'I'd have considered you, Sam', Susie said to me, 'if it hadn't been for your poor . . . you know . . .' I stared, uncomprehending. 'You know,' Susie was blushing, 'your . . . um . . . tentacles . . .'

When I returned to Hillman Street, on a stick, my young brother pointed to the empty space on the mantelpiece. 'I threw the tail fins away,' he said. 'Seeing as the whole north side of Belfast was feet deep in them, they weren't much good as novelties.'

By late 1941 McCleod, Burgess and I had had our aircrew selection boards. Mine was held in Weston-super-Mare.

'If you were flying an aeroplane in which the control wires had been accidentally crossed to give opposite effects, how would you go into a left-hand spiral dive?' I was asked at the outset.

'I'd pull the stick back and to the right with all sorts of right rudder,' I told the grey-haired officer, 'but thank God I won't be asked to do any such thing. I'm going for a navigator.'

'Are you really?' This guy consulted a file in front of him. Then he cleared his throat. 'Why don't you want to be a pilot? Are you afraid of the responsibility?'

'No more than yourself, sir,' I came back. He wore medal ribbons and an observer's badge. From the first war, by the colour of his hair.

The officer smiled. 'I'm not making much of a success of this, am I?' he said. 'This is my first selection board.' He wore squadron leader's rings. The junior officers on either side of him were looking uncomfortable.

'Well, sure, it's my first one too,' I said, 'and if I get to chair a board some day like you, I'll think I'm some kid.'

'I see you haven't had a secondary education.' A younger man had taken over.

'It would be more accurate to say that I haven't had any schooling at all,' I said. 'There is no record of any pupil at our school ever having passed an exam of any importance.'

'How would you cope with the mathematics of navigation?' he asked. I went into detail about the classes I was attending at Cranfield.

Then it was the third officer's turn. 'On the matter of discipline,' he said, 'it is noted here on your record that your engineering officer once had doubts on that score, but they are now eased. What caused the doubts in the first place?'

I took a very deep breath and, for the first time, paused before answering. This question was a stinker. Somehow I guessed that it was crucial: all three members of the board were watching me carefully. On my record before them, they could see that I had served over forty days' jankers, meaning confinement to barracks, for various misdemeanours.

'With the greatest possible respect, gentlemen,' I said, looking the chairman straight in the eye and offering up a squillion-volt prayer at the same time, 'when a senior officer who has known me for over a year expresses confidence in me as of now, does that not render the past unimportant?'

The squadron leader suddenly laughed. 'You're a cheeky bugger, McAughtry, aren't you? Is this why you've run up all those jankers?'

'More or less, sir,' I admitted.

Next morning I heard my name read out as one of the successful ones, and there weren't all that many. The squadron leader old-timer had one last word for me, as we passed in the main hall: 'What'll they say about you back in Belfast now?' He looked genuinely amused.

'They're already saying I was stupid for not staying in the aircraft factory,' I told him. 'If I get the hammer in this war it'll just go to prove it.'

The older man adjusted his hat before going out to his staff car. 'McAughtry,' he said, 'people like you never get the hammer in wars.'

So, back at Cranfield in the autumn of 1941, there we were, the three of us, still working as mechanics, but with our eyes on higher things. Burgess and McCleod had sailed through their interviews. They held their breaths as I told them about mine.

'I'd never have thought of that answer to the discipline thing,' Alec said, and Jack nodded agreement.

'But the thing is, it didn't arise in your cases,' I told them, 'because you both look the part. I look lippy. That's why I get into these fixes.'

Jack Burgess, smaller and more finely built than myself, was intelligent and looked it. His wit was mercurial, his grasp of facts was sure and quick. If he'd had McCleod's opportunities he'd have been honours degree material. Come to that, if McCleod hadn't been so lazy, he'd have been a flight lieutenant navigator already. Just because he liked the company of people like Burgess and myself he'd hung around, living like a pig in barrack rooms and tents, taking his time before starting his flying career because he couldn't be bothered doing anything else. He looked like a figure out of Stevenson's *Kidnapped* – reddish hair, strong, handsome face, powerful, tall build. And that Highland tongue – as much at ease with Gaelic as with English – I could listen to it all day, only I wasn't going to. We were breaking up. We felt it, without saying anything.

The play-acting was over now. None of us could afford to get into any sort of trouble, so close to aircrew posting. In the evenings we drank as quietly and circumspectly as the ones we'd once despised. Often we didn't go out at all, but sat around the barrack-room table, helping each other with flying subjects. A few months earlier, if we'd had a similar group acting that way in our billet we'd have tormented the hell out of the poor bastards, but with our reputation nobody tried the same on us.

The day came when Alec and I stood at the main gate and reached up to the tail of a three-tonner to shake hands with Burgess. He was going to Stratford-upon-Avon to commence initial pilot's training. 'We might never meet again,' he said, melodramatically. It was September 1941.

A month later Alec's turn came. He was going to London. His farewell at the guardroom began with his finger pointing to it. 'We've seen the inside of that place more than anybody in this camp, anyway,' he said, 'so I suppose we'll be remembered. It's one kind of memorial.'

'Are you sorry our frigging-about days are over?' I asked.

He nodded. 'It's all thinking from now on, isn't it?'

We shook hands. His final words to me were in Scots Gaelic. 'Is that good?' I asked. He nodded and got onto the truck, still nodding. We were bloody nearly crying, the pair of us. Some men-at-arms, I must say.

When I got to the Aircrew Receiving Centre in St John's Wood, London it was December 1941. Alec McCleod and Jack Burgess had long since passed through the place. I was the only serving airman in the entry of forty-odd, the others being mostly fresh-faced college and public-school boys. I must have looked like a real hard lick to the new boys, with my propellers of rank on my sleeve, buttons rubbed smooth, and my forage cap worn so far to the side that I seemed to be walking alongside it.

I was the senior cadet. This made it possible for me to remove an obstacle from my path. Some months earlier I had fallen out with a tall, rawboned flight mechanic from Durham, and the two of us had fought to a standstill down on the flight. I split his eye, but he broke my nose, and the medical officer at Cranfield sent me for examination to an RAF hospital at Rauceby in Lincolnshire.

There, a cold-eyed squadron leader prodded and tested my poor nose. 'We'll have to have you in here,' he said.

'I'm not surprised,' says I, willing enough at that stage.

'Oh, yes,' he said, 'we'll have to fillet that bone out.'

'You'll have to do what?'

'We'll fillet that bone out.'

I rose, reached for my cap and threw my small kitbag and gas mask over my shoulder. Then I saluted. 'Good morning, sir. Thanks for your time.'

'Where the devil do you think you're going?'

'I'm going back to camp. And I'm taking my nose bone with me. Many thanks.'

'Do you refuse this operation?'

'Vehemently. Flatly. Uncompromisingly. With every nerve in my system.'

'Then wait . . .' He turned to his desk, wrote quickly on an RAF form, put the form into an envelope, licked it, and handed it to me. I stood at ease and put my hands behind my back as the squadron leader delivered a final rebuke. Listening to him tone down, trying to soften the wording, attempting to persuade me one final time to have the operation, I opened the flap of the envelope. Refusing for the final time, I was dismissed. Outside, I read the report:

'The airway is completely obstructed. This airman flatly refuses to undergo an operation.'

The medical officer at Cranfield was Irish. Sportingly, he went along with my plea to let the authorities in London resolve the matter.

One of the first parades of the new aircrew entry was to the orderly room, to collect individual Form 48s – the medical records which accompany airmen wherever they go in the service. As senior cadet, I was given the forms for distribution. I took my own, went into the jakes and read the file copy of my report by the doctor at Rauceby. Then I removed it from the file, tore it to pieces, and flushed it down the drain.

My duties also meant that I worked closely with the drill corporal in charge of our intake. On the very first morning I noticed that he marched us the guts of half a mile to a little teashop, for break. He did the same in the afternoon.

I spoke to the corporal. 'Forty cups of tea and forty buns, twice a day, five days a week. That's four hundred cups of tea and four hundred wads. The shopkeeper's very fond of you and me over the head of that, isn't she? She's not going to see you and me stuck for a packet of smokes and the price of a pint, now, is she?'

'And what would you call the price of a pint, now, Paddy?'

'Just whatever you think my influence is worth in this outfit. And I'm worshipped in the ranks, mind. These boys have never been on an airfield any more than yourself, and I'm reeking with the smell of hundred octane. I have my eye on a wee shop not a hundred yards from the billets. I might lead them there. You never know.'

The shopkeeper was lifting £13 a week, making probably £10. We ended up with thirty bob each, the corporal and I, and the best of drinking pals we became too.

At this place we were given medical and intelligence tests. We were also tested in proficiency at the Morse code. I wasn't prepared for this. In a panic I rushed out to Woolworths and bought a book called *Teach Yourself the Morse Code*. I had only a week to master the thing before the test, and if the test was failed it could mean total failure at the outset.

The book's system was very good. For each Morse symbol a spoken phrase with the same sound was substituted; for example, the letter Y is 'dah dee dah dah'. The phrase to help remember it was 'You can do yours'. Very ingenious. Everywhere I went I was muttering 'You can do yours' and various other strange-sounding sentences. I got some very queer looks, but the thing had to be done, and that was all about it.

If Y was 'You can do yours', then Q was 'God save the Queen'. The two sounds are very similar. 'God save the Queen. You can do yours,' I said to the cook, as he ladled bangers and mash onto my plate.

'Bloody Irishmen,' he muttered. 'What's he joined up for if he don't like England?'

The time came to take the test. I was a bundle of nerves. Lots of the kids in the class had done Morse in the Air Cadets and the Boy Scouts. I'd never in my life heard a signal buzzer. It would be a daft way to fail as a navigator, after all my trouble.

'Right, chaps,' this officer said, 'I'm going to send at four words a minute to separate the wheat from the chaff. Three minutes' transmission, starting now.'

It was money for jam. I was delighted. The buzzer was actually talking to me in a new language. 'Chase me Charlie', it burbled, and I wrote down the letter C.

'X marks the spot', my little friend said, and I noted the letter X. Beside me some fool was under the impression that L marked the spot. I was eager for the next symbol.

When the operator had finished I was full of confidence. Sure enough, I'd achieved 100 per cent. 'Now we're stepping it up to six words a minute,' the officer said, bending to his work.

'Send away there,' says I to myself.

'Freddie's freezing', came the message. 'Oh ho,' says I, 'so he is indeed,' and down went F on my sheet. 'In London town', the contraption said, and an L joined the F. Another 100 per cent and no bother, and not only that, but there were now only four of us in the hunt.

'Eight words a minute now,' the officer said, excitedly, and I was thinking to myself that I could end up being excused Morse code for the rest of my training, I was so brilliant at it. At the end of the test I had scored ninety per cent, the highest mark of the lot. 'And mind you,' says I, all euphoric, 'I only started learning it a couple of weeks ago.'

'I want to speak to you privately,' the officer said. Full of pride, I winked at the grammar-school boys as they filed out. I was pretty sure I was going to be asked to stay on in St John's Wood as an instructor. I had my answer ready. 'No thank you,' I was going to say, firmly. 'I joined up to do a job and a job I'm going to do.' That's what I was going to tell the officer.

'We need picked people for a special job,' said the officer, 'and you, Paddy, fill the bill beautifully.' My eyes began to resemble slits.

'We simply can't find enough men with the ability to operate the radio transmitter, navigate the aeroplane and fire the Browning,' said your man, 'so you can see how happy you have made me today. This is a very special day in your life,' he went on. 'You have been earmarked for special operations.'

'In the course of that peroration,' I said, 'you have used the word "special" three times. Why should that fill me with foreboding?'

I remembered the words of Uncle Alec: volunteer for nothing.

'For your information, I have been fooling the RAF,' I said. 'I can't really do the Morse code. All I've been doing is saying "You can do yours" to myself, and silly things like "God save the Queen".'

'Well, whatever it is,' said the officer happily, 'it'll do me. People with your gifts are becoming scarcer by the hour.'

'Oh, is that a fact?' I said. 'And why do you think that is?'

'They're being shot down like fucking flies, that's why,' was the reply.

'Doing exactly what?' My eyes were shut tight. I was having

difficulty in swallowing.

'Flying in Bristol Beaufighter aircraft, in daylight, attacking enemy harbours and shipping convoys, that's what. Do you know,' he went on, warming to the subject, 'when bomber aircrew try to avoid operations through cowardice, they are often sent to Beaufighter squadrons as an unofficial death sentence. Isn't that poetic justice?'

'I wish to see the commanding officer,' I said. 'Low-level daylight flying gives me sore ears. Anyway, my airway is completely obstructed. Look.'

The officer patted me fondly on the back. 'Go and pack your kit, Paddy,' he said, 'and for the rest of your training, look around you. Savour the flowers, the birdsong. Retain the tang of good bitter beer. Linger with English girls a little longer than is usual. It won't be all that long now till you kiss goodbye to the whole frigging lot.'

Somebody told me in the canteen that night that the officer's father had been shot out of a Crossley tender in Dublin. He must have been out for revenge. I should have made it clear from the outset that I was a Protestant Loyalist, reared in the orange and purple.

4

I become a spark

During lectures at St John's Wood I'd found that my maths were very shaky. I think I was the worst in the class in this area. When I went to start initial training at Babbacombe, near Torquay in Devon, I got down to some hefty boning, but at the fourth-week test I failed to reach the minimum mark. Luckily this wasn't a crucial matter, but if I were to do the same at the eighth-week test I would be out on my ass, and would have to try air gunnery instead. An air gunner's life was too boring to be contemplated.

I don't think I have ever worked as hard at anything in my life. These two fishermen wanted to cross a river in a boat that only did five knots, whereas the current in the river was ten knots, and so what was their compass heading, if they were to nose eventually into that landing stage two miles downriver? Then there was the matter of compasses and magnetism. I'd never in my life thought there was anything complicated about compasses, but the thing turned out to be plagued by variations, deviations and coefficients.

The Morse went on relentlessly. We weren't to be ordinary observers, with just enough working wireless knowledge to get by – we had to be pukka air operators, as well as having all the other skills.

There was a heat wave that year. It meant nothing to me. I was indoors most of the time, reading. The cream of the Luftwaffe, in their yellow-nosed Messerschmitts, strafed the main street of Torquay one day, and there I was in the middle of it. It could have been Sir Alan Cobham's Flying Circus for all the difference it made to me. I was thinking of sines and cosines and tangents at the time.

I was sitting on a hilltop with a Scot called Angus Duncan when two London munitions factory girls sat down nearby. The sun was straight overhead in a cloudless sky. Angus began to work on the girls. They were our own age, about twenty. I couldn't be bothered. Neither could they. 'We're down here for a rest,' they said, 'so piss off.'

Well, we didn't piss off. Instead, I fell asleep with my head on the lap of one of the girls. The other girl slept beside us, and poor Angus, who was too brainy to be tired studying, chafed and fretted the whole afternoon, unable to believe what was happening. Still, we all made up for our lassitude that night. The girls were newly arrived. We got to date them first, before about three hundred aircrew cadets. It was like winning the Irish Sweep.

I had one serious aberration at Babbacombe. One night, when I was supposed to be confined to the hotel on fire picket, I shoved my kitbag under my bedclothes, went out and got numb-in-the-legs drunk. The drink was cider. It was sheer dynamite. But there was good reason for the risk I was taking. The sergeant who taught us anti-gas and decontamination measures was known to be a wild cider drinker, who ran out of boozing money two days after payday.

Running the sergeant to earth, I began to ply him with drink, sinking a fair old ration myself. Final exam time was close. 'Is it always the same exam questions?' I asked, cunningly.

'Always the same,' he said.

'So, if a big drink-buyer was about the place, he could maybe find the whole exam paper on the pub table some night,' I suggested.

'You don't have to put it off till some night,' the sergeant said. 'See me all right tonight and see me back to my billet, and you can take a copy of the bloody thing back with you.'

Which is what I did. Only when I got back the place was in an uproar. The orderly officer was a keen type, an ex-schoolmaster, who wanted to visit the hotel billet and brief the fire picket during his spell of duty. All he found to brief in my room was my kitbag.

I was brought before him, next door to poleaxed. I remember that, and nothing else.

'God take care of me,' I said to Angus Duncan when I woke up next morning, 'but my brain tells me I'm up to the neck in it.'

'That you are,' he said. He was a nice guy, and sorry for me, but not the sort to get into that kind of fix himself. Already he was distancing himself. I could see it in his manner. As I was making up my bed a runner came to tell me to report to the officer, whose name was Kissock, known unfairly as Kiss-Me-Arse. 'Heavens,'

I groaned, 'a bollocking before breakfast. The man's a sadist.'

In fact he was anything but. 'Do you know, Leading Aircraftman,' he began, 'I've always been an abstemious man. Drinking to excess fills me with revulsion. You were hardly able to stand last night.'

I nearly told him the reason, and almost showed him the exam questions for proof. That's how shot to pieces I was. But I said nothing, just stood rigidly to attention, a sight that made him shudder, as it did everyone else who liked smartness in bearing.

'I'm still surprised at myself for my leniency last night . . .'

The hungover heart gave a surge. Leniency?

' . . . But your remorse was so genuine, and your worry about the effect on your father of your expulsion from this course was so intense, that I promised you last night that I would sleep on it.'

He was a forty-five-year-old pilot officer, the lowest rank for an officer. His chest was bare of medals or wings. He was slightly built, with rimless glasses, and his officer's ceremonial hat sat on his head dead straight, without a single dent, without the flying man's rakish slant or bent peak. Kissock was a thoroughly nice little man who was trying to help in the war as best he could.

'You come from a depressed background, yet despite that, you have reached this level in the service,' he said. 'Your parents must have been very proud when your brother got the Military Cross in the Army . . .'

Sweet Saviour, says I. I must have been good last night. Military Cross. Nice one, that. And the old man upset if I was humped off training? Sure he expected nothing else from me.

'Against all my judgment, I have decided to overlook your serious breach. You will do a week's fire picket. Dismiss.'

'The best way I can pay you back is in the quality of my work, sir,' I said, earnestly. 'Just you make a point of checking on my next exam results, that's all.'

'Which subject is next?' he asked.

'Anti-gas and decontamination,' I said. 'I've been swotting up on it like anything.'

Cranwell, in Lincolnshire, is the Sandhurst of the RAF, although the peacetime-style training of officers had been suspended when I went there, in mid-1942. Once a naval airfield in the first war, where Sopwith Pups and Farmans sputtered over the flat fields, it became an RAF station when the Navy and the Royal Flying Corps amalgamated to form the new service in 1918. Hugh Trenchard, the father of the Air Force, took steps to put the RAF on an equal footing with the other two services in 1920, when he got agreement to locate the first RAF college there.

Although my purpose in going to Cranwell was to train as a radio operator, nevertheless the tradition of the place was so strong that I and the other cadet observers became aware from the outset that we were at the very heart of the Royal Air Force: the place where the first winner of the Schneider Trophy, in 1929, Flying Officer Henry Waghorn, had been trained; where the legendary Douglas Bader, the legless pilot, had been a cadet; where Peter Townsend, the Battle of Britain ace, had learned to fly.

The RAF College itself is an imposing building, with pillared entrance, wide and gracious in appearance. Above the main entrance was a high tower, but soon after I arrived a Polish pilot flew a Whitley bomber clean into the tower, demolishing it and killing all on board. The wreckage clung to the spot for days, reminding every cadet on the station of the risks of flying itself, never mind the threat from the enemy.

Nearby was Sleaford, a small town, so much a part of the Cranwell garrison that the local innkeepers and business people had long years before forgotten how to greet visitors in friendly fashion. When I went drinking in Sleaford, in a pub called the Marquis of Granby, I always begrudged the town the money. It had no character, even the men who lived there behaved like transients, in a hurry to clear off somewhere else.

On the ground, for hour after boring hour, I sat in classrooms listening to automatic Morse machines churning out plain language and code groups, but I was good at it, so it didn't get me down as much as it did the others. The technical side of radio was a different kettle of fish, however. We had to learn how the transmitters and receivers actually worked. This involved mastering the workings of such as the diode triode push-pull valve, and making ourselves familiar with master oscillators and the like. We were taught that the low-tension battery heated the thing that sat at the bottom of the valve, and that this made it easy for the high-tension stuff to leap smartly from the bottom thing to the top thing. And if things didn't work they nearly always would if you kicked them.

We flew in Percival Proctors and de Havilland Dominies, setting up our transmitters and calling base. The first time that I made two-way contact and actually began to pass messages over the air gave me the most wonderful feeling. Indeed, it was something I never lost, the magic of talking by Morse key to someone on the ground many miles away, someone whose whole purpose in life was to be friendly and to help.

For practical purposes I was never really a pupil at Cranwell. Wireless operators were like writers, in that no two of them used the same style, and all of them loved good workmanship. I took to the practical, operating side like a duck to water, and my shortcomings on the theoretical side didn't matter, for there were simple fault-finding tips that covered most eventualities in the air.

The Dominie was a biplane. In the air its every rivet, brace and strut quivered. The slightest change in air pressure could be felt through the machine from nose to tail. To fly in the canvas-covered biplane was to experience real flying. I couldn't get enough of it, and the radio instructors, seeing rare enthusiasm, brought me into membership of their club. I had a distinctive sending style, and was more often than not greeted enthusiastically over the air by permanent staff operators delighted to hear a different, interesting 'fist', as they called it.

'GMOM' they would send, when two-way contact had been established. This meant 'Good morning, old man', and when I first had it addressed to me it charmed and made golden my whole day, for the greeting was confined to the initiated. In the evening, in a dark sky, returning home on radio-loop bearings, it was gorgeous to hear the Cranwell ground station signing off with 'Good night, old man'. The ground operators knew that we were essentially navigators, that radio was only a second string, that almost all the pupils would forget half of their teaching as soon as they left Cranwell. We were a queer breed. Normally in bomber aircraft there was a wireless operator who did nothing else. But we were going to fly in two-seater Bristol Beaufighters. There were only the pilot and the navigator, so the latter had to be a Jack-of-all-trades. In the air, whilst transmitting, Beaufighter navigators were generally dreadfully bad, and they got worse the longer they flew, in my experience. But I revelled in this side of the job, and many a time it got me out of a fix, too, in the days to come.

Amazingly, we had to learn semaphore. We were taught this, together with Aldis-lamp signalling, by two darling old characters called Wink and Wank. It was Wink who taught semaphore. I was good at this too, for old Wink's sake. He was a retired naval petty officer yeoman of signals. His battleship days were done, and the pupils made fun of him. But I didn't. I displayed my flags crisply and at tidy, accurate angles. 'Open fire' was one signal. 'You are standing into danger' was another. 'There you are, look at McAughtry's flags,' Wink would say, his pale eyes shining. 'If his aeroplane's ever down in the drink, he'll be able to send for help while the rest of you are slowly sinking.'

Because Wank was Wink's very good friend, and because both were very friendly with the man who taught air traffic control, I was called to one side in my last week at Cranwell and given the questions that would be asked in the air traffic control examination. I got 100 per cent in it. I also got 100 per cent for aircraft recognition, because a chap called Boardman and I agreed that I could cog aircraft recognition from him, if he could cog Morse from me. I passed out from Cranwell with the highest marks ever given to anybody by any examiners anywhere in the world.

It was at Cranwell that our entry sustained its first casualty. An engine appeared to fail during take-off and a Percival Proctor crashed on the edge of the airfield. The aircraft caught fire and the staff pilot and pupil, probably trapped in a twisted fuselage, were unable to get out. After the fire crew had smothered the flames with foam, the charred heads of both men could be seen, showing just over the top of each cockpit. I saw the wreck before the bodies were lifted out. With me were a good dozen from the dead pupil's course.

I was surprised to see how matter-of-factly the others took the loss of their comrade. I'd become well used to casualties on an RAF station and, indeed, in the Battle of Britain I had lost several friends on the ground staff, killed on coastal stations by bombing and strafing raiders, but the new boys, after some initial shock, simply followed armed forces' custom – dropped the dead boy's name from their conversation and carried on as before.

After only two weeks at Cranwell I was joined by a Derryman called Mark Kerr. He had just been washed out from a pilot's course. Not for any shortcoming in the air but because, for the moment, the Air Force had enough pupil pilots. The new, specialised post of bomb aimer had been created. They were needed in a hurry, so the RAF arbitrarily washed out a certain percentage of pupil pilots and offered them training as bomb aimers. It saved time putting the new men through initial training, in the short term, and indeed, in the beginning, bomb aimers were seen as emergency second pilots, who could, in a desperate pinch, get a bomber home and onto the ground, so the idea of starting off with men who had actually soloed on Tiger Moth trainers was an attractive one. Mark Kerr, though, thought that navigating offered more of a challenge.

At the same time I had a letter from Jack Burgess telling me that he too had been washed out. I rushed to send a telegram: 'Join me at once. There are ex-pilot trainees arriving here now. We can be together again.' I was quite excited, waiting for his reply. It took two months in coming. I was halfway through training at Cranwell when a wire came for me: 'Enjoying best leave of the war. Wishing you the same good fortune as has attended me. Pilot Officer J. Burgess.' He'd completed the bomb aimer's course and passed out among the top scorers. I remember reading and re-reading his telegram and wondering what the hell could have happened to his prose sense. As a rule, Burgess threw words together much better than that.

I passed out without any trouble at Cranwell, put up my wireless operator's badge on my right sleeve, and went home for a week's leave. My old dad was at home, working ashore on a ship that had been badly burnt in an air raid. He was enormously excited to see my new badge: 'A spark, eh?' It was the first time in my life I had ever seen him impressed by anything. The wireless operator, or spark, on board ship ranks as a junior officer. My dad seemed, for the first time, to see some serious purpose in the training I was taking in the Air Force. He took me round to the Spa Inn, showed me off to all his pals, and pointed to the badge: 'He's a spark, so he is.'

I remember groaning to myself. Really, that book I'd bought in London had taught me all I needed to know about operating. The rest was just drudgery. Neither Dad nor anybody else in our family had a bloody clue as to the size of the job ahead of me in mastering air navigation, mainly over the sea at that. When I mentioned this to Dad, he just shrugged. 'That's just dead reckoning and that sort of stuff, isn't it? Some of the stupidest bastards I've ever met can do that at sea. But sending SOSs and the like of that? It's a job takes some doing.'

Mark Kerr travelled with me on that leave. The Heysham boat docked at seven in the morning in Belfast. This created a problem which had been exercising my mind ever since we'd left Cranwell with our kitbags.

In order to get to Derry, Mark would have to catch a train from York Street station, only a few hundred yards from my home. I worried about the time of that train's departure all the way across on the boat. Eventually, as we were getting ready to disembark at Belfast, my worst fears were confirmed. 'That bloody train of mine doesn't leave till twelve noon,' Mark said.

My heart sank. I was faced with it. I ought to invite him to Hillman Street, to pass the next five hours.

'There's not a bad canteen on the station,' I said. 'Get a mug of tea inside you, and I'll join you by nine. We'll take a walk and I'll show you my end of the town.' And that's the way I left him.

I felt bloody awful, drinking tea with him first, then Guinness, when the pubs opened at ten. I saw Mark onto the Derry train and told him I'd meet him at the boat a week later. After only one day I could no longer live with myself. I sent a postcard to Shantallow, in Derry: 'Am bored. Come on down for a couple of days.'

He arrived two days later. We went out on the town, had a wonderful time, drank in the Spa Inn with my old dad, showed everybody in the bar Mark's wireless operator's flash, went to the dog racing with my younger brother, looked up all my mates, drank tea in their homes and sat on butter boxes in some of them instead of chairs. Mark Kerr slept on our sofa. He had to go home for the last couple of days. 'See you at the boat,' I said, smiling all over my face. Feeling great.

After that, I was never again ashamed of our house, with its outside WC, and its bath in the attic that needed the hot water brought up to it in buckets, and the dog shit in the back entry. And I was never ashamed of my mother again, either, or my old dad, who rated an RAF wireless operator equal with the first mate of a ship. Or of my very good friends in the neighbourhood, who were tongue-tied in Mark's presence, till they saw that I was his mate, so it followed that they must be too. But I remember thinking also, as Mark waved from the train, that I'd just had a near miss from some kind of disaster or other. Definitely.

At the end of October 1942 we were taken from Heaton Park in Manchester to the Tail of the Bank in Gourock, Glasgow. It was all very mysterious. In the dock we were led along a quayside, until we came to the gangway of a passenger liner. This ship's upperworks were higher than any I'd ever imagined, let alone seen. On either side its hull stretched into the gloom. Neither its bows nor its stern were visible, they were so far apart, and we were amidships. It was so big that I had to stop and stare. It took the shoving of the cadets behind to get me started up the gangway.

It was the RMS *Queen Elizabeth*, at 80,000 tons the biggest ship in the world. We were on our way to Canada, via New York. We'd already been blessed by the RAF chaplain in Manchester, in case we were torpedoed.

When I went aboard the liner *Queen Elizabeth* to go to the States I left behind a Britain where, because of the crippling losses of ships in the Atlantic, the natives were beginning to regard horse meat as a delicacy, where it was routine to sit down to a breakfast of powdered egg and grey-coloured bread washed down with tea lightened with milk powder, with no sugar and no handle on the cup. Back home I had two brothers who stopped talking to each other for nine months because the one came home to Hillman Street unexpectedly to find the other sitting down to a secret feed of black-market bacon and fresh eggs. I went down to my cabin in the *QE* with, tucked in my wallet, a written promise from a twenty-year-old wireless operator in the Women's Auxiliary Air Force to the effect that she would undertake to do 'anything, anything at all' that I wanted her to do, if I would only bring back for her two pairs of nylon stockings. The ladies had resorted to dyeing their legs a light tan and drawing a line with black pencil up the backs from heel to thigh, to simulate seams. The thing is, the lady wireless operator had had the chance to do 'anything, anything at all' for me already, on account of my bringing back for her family, who lived in Sleaford near the RAF station at Cranwell, a sizeable pig's knee, obtained through the good offices of my Uncle Sammy, who was a knife man in Sinclair's bacon factory in Belfast. But all I got was a kiss. Still, I was looking forward to more slavering from stocking-starved females, when I got back from flying training in Canada. More immediately, I was looking forward to some decent grub for the first time in two years. On the *Queen Elizabeth*, victualled in New York, the change in the quality of food was so sharp and sudden that it actually weakened me initially, in much the same way that too much fresh water puts the thirst-crazed desert traveller in peril.

We dined in the huge ship's restaurants, and were waited upon by stewards. It was possible to see the flesh growing on us. After rising with difficulty from a breakfast of orange juice, cornflakes with fresh milk, ham and three eggs, with tea and sugar and more fresh milk drunk from a cup with a handle, I used to cram further rashers of bacon into my pocket and chew them through the morning until it was time for lunch. From the luncheon table I took meat, and chewed it through till teatime. All along the space behind my bunk I had a row of hard-boiled eggs, and I ate these as I played poker, which was all the time between sleeping and eating.

The actor Edward G. Robinson was on board. He was fond of talking to the troops. I told him once how the sudden change to good grub had affected me. He stroked his chin. 'Amazing,' he said.

'That's just the way you said it in *The Amazing Dr Clitterhouse*,' I said, and he hugged me and gave me a cigar.

In the bows of the *Queen Elizabeth* a series of poker and dice games went on right round the clock. The players were airmen like myself, plus Yanks going back Stateside, and Army men who were part of the permanent staff. But outnumbering all these were crewmen from the *Queen Elizabeth* herself. How they got away with it was a mystery. They were deck hands for the most part and, watching them, I remember hoping and even praying that the lookout wasn't one of the guys hugging cards to his chest.

The door to the room in which the games went on was locked, and a doorman was paid by the players to peep through a window and identify visitors. The ship was crawling with Army NCOs searching for victims to do fatigue duty in the ship's kitchen. It was my proud boast that I was not nailed for duty once on the four-and-a-half-day voyage. When a duty NCO approached the room where we were playing poker and crap he either found the door locked against him and turned away or, if he persisted, he was given money and turned away just the same.

The seas became dog-rough once we got out properly into the western ocean, and there was a more or less constant trickle of men staggering weakly from the games to the heads. Back in the cabins others were unable to leave their bunks. Whereas I must have added half a stone to my weight during the voyage, others, like poor Mark Kerr, looked near to death when we were halfway across the Atlantic. Mark was also a victim to airsickness, yet he completed his whole flying programme, and was awarded his flying badge alongside myself. He deserved the Victoria Cross. I deserved what I got on the *Queen Elizabeth* too. I made over a hundred quid at the poker. I've never in my life seen worse poker players, especially the Yank soldiers, who were portrayed in the cinema as such cigar-smoking poker terrors. Any kid from the street where I grew up in Belfast could have taken the pants off the Americans, and a funny thing is that I also found their drinking to be a lot less iron-headed than we'd been led to expect. As far as rye whiskey was concerned, I could have carried more in my hollow tooth. Hollywood did the average American male a disservice. The discovery that they were just like ourselves was one of my biggest disappointments.

The sight of the New York skyline was a marvellous thrill, especially from the deck of the *Queen Elizabeth*, and this arrival from the British war zone was another occasion for the Americans to show their friendliness, by way of fire-boat demonstrations and sirens sounding as we passed the Statue of Liberty, heading up to our berth. Yet, oddly, the sight that impressed me most was the traffic making its way over the Hudson Bridge. There wasn't much private motoring going on back home, and such cars and trucks as were on the roads were camouflaged in dark green or, if private, tended to be dark blue or black in colour. Here were hundreds of brightly-coloured cars tearing along, with yellow cabs predominant. This was to become commonplace at home in postwar times but, seeing New York traffic for that first time, I felt as though I was watching an exciting carnival, a colourful glimpse of fairyland, where no danger threatened, where any nice thing in the world could happen.

In the event, when I got down to New York later to look the place over, although nice things did happen, I took an unmerciful thump across the side of the jaw as well, from an American navyman, which just goes to show the sort of rubbish that went through my head, when I stood, like all the bog-Irish before me, and looked for the first time at the New World.

Once established at the Royal Canadian Air Force station at Mount Hope, Ontario, I had a lovely surprise. I was placed on the payscale of an RCAF leading aircraftman flight rigger, since this was my basic trade, and the money certainly was nice. It was actually possible to go out into Hamilton, the nearest town, do the Champagne Charlie stuff with a woman three times a week, and still have money left on payday. Another pleasant surprise was that I no longer had to work so hard in the evenings just to keep up with the rest. I had the knack of air navigation once we began to fly, although, after my first trip in an Avro Anson across a large slice of Ontario, I thought I was washed out, a candidate for the next train to bomb-aiming school at Toronto for sure.

We'd had lots of classroom stuff, tons of theory. The time came to get our feet off the ground and put it into practice. We were given the forecast wind and off we went. My pilot was a serious-faced Englishman, Pilot Officer Fletcher. As soon as we were airborne

I gave three quick changes of course to check the forecast wind. My figures showed it to be wildly out, so I gave a new heading for the first leg. After five minutes the pilot made my head shoot up from my chart table by asking why we were crossing the coast of Lake Ontario and heading for Alaska. Hurriedly I checked the wind again, got another value and got the aeroplane back, heading, as I thought, for our first turning point. It struck me that it mightn't be a bad idea to check my calculations by having a radio-loop bearing taken, so I got the radio man to whirl the loop round and let me have the findings. This showed my last forecast wind was a lot of rubbish so, fair enough, I said, you'll just have to alter course again, pilot.

We staggered around that triangular course like a wine victim trying to make it to Alcoholics Anonymous. On each of the three legs I made a good half-dozen alterations. By the time we made it back to Mount Hope the crew were exhausted and the airfield authorities were thinking of sending out a search plane. And just to rub my nose in it, the wind that brought me home finally was the one I'd been given in the first place.

Next day in the classroom I found to my horror that, because of the accurate wind forecast, every other flier on the course had had an armchair ride, turning on time and landing within minutes of their forecast times of arrival.

The squadron leader instructor read each man's flying log details out but, surprisingly, awarded very low marks in each case. He kept mine to the last. Holding it up to view, as I shrank down behind my desk, he said, 'Take a look at this effort. Note the sweat stains. Observe that his entries marking the wild progress of this flight cover all of one side of the log and most of the reverse. Here's a man who doesn't believe our expert meteorologists and sets out to find his own wind. He flies half over the continent and eventually gets back to base with his crew nervous wrecks.'

'Stand up,' the squadron leader ordered.

'Just throw me off the frigging course and leave it at that,' I growled, mortified. 'Don't be making a mug out of me altogether.'

'Stand up,' came the order again.

Glowering, I got to my feet, marking carefully for future

reference the faces of those ready to titter loudest.

'He's the only bloody navigator on the course,' the instructor said. 'He's the only one who doesn't believe forecasts. He has made more mistakes and got himself into more cock-ups than the rest of you put together and multiplied by ten. And what's more important, he's learned how to get himself out of them. There's almost nothing more I can teach this one about recovering from errors, and he's got ten out of ten to show for it.'

Well, of course, it was unwise of the squadron leader to blow my head up like that. As and from that day, I did the minimum amount of work, did all the night spots in Hamilton and Toronto, climbed the airfield fence at all hours, and discovered Gladys.

5

The air observer

At twenty-one the young male eye is sharp and quick to see signs of added years in a woman, and Gladys had oodles of them. The skin wasn't just clear enough, the eyes a little too wise. Gladys had fair hair. She was slim, stood five inches below my five-eleven. The lady was full of zest, but another thing that placed her near to the thirty mark was the natural way she had in the Canadian winter with muffs and furs. Girls in their teens and early twenties wearing furs always look as though they are in clothes borrowed from Mummy.

I met her in the Connaught Rooms in Hamilton. I was drinking with the corporal cook and Gladys was with dark, plain Meta, another thirty-year-old, four years over from Rochdale in Lancashire, half Barbara Stanwyck in the voice and half Gracie Fields. Gladys was Canadian born, slightly Indian in appearance. Her voice was husky and her wit was dry.

The corporal cook hadn't told me that he was dating Meta, and that I was the one for Gladys that evening. He just asked me if I'd like a drink when I happened to come into the bar. When the two ladies turned up I made a pass and it took me a good three minutes grabbing a table and organising beers before it dawned on me that the cook was the main man from the start.

I told Gladys that I was the sole heir to an Irish shipping line and she came right back and said that, yeah, she knew how it was, on account of she stood to get the majority shareholding in Eaton's Stores when the old man died. Some aircrew cadet let his glance rest on the women at our table for longer than the permitted millisecond and I asked him if he'd like his seeing eye closed for the night. Then I explained to Gladys that I'd have to be very careful with my protective instincts because, as a professional welterweight boxer, my hands were regarded by the courts as lethal weapons. On hearing this she looked at my hands and said that they reminded her more of the hands of a man who sold women's clothes. Which was very funny, as I told her, in more relaxed mood, because that's exactly what I did once, when I was sixteen.

We had a nice, talkative evening. After the hotel we went to Meta's flat nearby and the furs went across the backs of chairs. Almost at once Meta and the corporal cook went through a door and disappeared from human ken. Gladys made tea for herself, fixed a beer for me and then we sat on the settee. She had slipped off the jacket of her suit, and sat relaxing in skirt and woolly jumper, with her snow boots off, and her nylon tops rolled around a garter above her knee, a spot beyond which I could not go, not that I really expected to.

I finished off the three beers left in the icebox, told Gladys to stop biting my lip when she kissed me, made a token effort now and again to climb the north face of her thigh, but, being philosophical about my sexual pulling-power, felt very happy with the way things were turning out.

Then I had to go to the bathroom, by way of the door that had swallowed the other two. I went through, and came at once upon a double bed, and in it, with their two heads together, quiet and sleepy and satisfied, were Meta and the corporal cook. As their eyes turned to me it suddenly dawned on me that he spoke Rochdale too. This was the sort of coincidence it took to land a man into bed with a woman. I'd only ever seen one other couple in bed before in all my life and that was my mother and father. In the bathroom I remember feeling very young and very green. There was only one thing to do about such a feeling. 'After you two with the scratcher,' I said, on the way past.

'You cheeky young bugger,' Meta said. In her outrage her voice was as Rochdale as a mill bobbin.

'Hey, does this sofa open out into a bed?' I wanted to know, when I reached Gladys. She just took a hold of me and started back into the hugging and kissing. It suited me well enough, who was never used to much else, with the exception of Martha, back in Bedfordshire. As I dropped back into the role that had been chosen for me, I kept trying to remember the other advantage that the corporal cook had over me. Screwing up my face against another of the deadly Ontario lip-bites, it came to me. He was thirty as well. I was going to have a long wait for the consummation, but things could have been worse. When I'd walked into the Connaught Rooms that night I'd no woman at all. Now I had right of way over the top and bottom thirds of a counter hand in Eaton's Stores.

In the early hours of the morning the corporal cook and I walked back to camp and climbed the wire. He wasn't the talkative sort. The only thing he said was that he was going back to England that very day. I couldn't wait to see Angus Duncan.

'I'll sell you a woman who's been sleeping with the corporal cook,' I offered. The deal was struck for an RAF-crested, white, high-necked pullover, and I rang Gladys and told her that I had a friend for Meta the following Friday night.

We all went to the bowling alley on the night and played bowls and drank beer for the million years before it was time to go to the flat. If Angus gets it I'll get it, I kept telling myself, excitedly.

We all sat down, Gladys on my knee on the settee, Meta on Angus Duncan's on the easy chair. We nibbled and beered for a bit, then I caught Duncan's eye and nodded in a meaningful way to the door that led to the bedroom. Angus suddenly rose, his arms around Meta's shoulders and legs, and attempted to carry her over the threshold into the bedroom.

Gladys burst out laughing at the sight. Duncan was really quite small and the effort was clearly herculean for him. After an initial startled yelp Meta struggled free, and stood fixing her straps and hitching her nylons back up.

'What wirr ye auchtermuchty heelan laddie wi the corporal cook and why no wi me!' Duncan spluttered, or words to that effect.

'Tha'rt reet savage,' Meta yelled, throwing him out onto the landing. 'Do thee think I'm Mademoiselle from Armentières? Tha'rt worse savage. Were thee started it,' she shouted at me, and out I went too.

It was too late for the bus. That night Angus Duncan and I slept in the gymnasium of the Hamilton YMCA. Every so often, whenever I thought back to the scene in the flat, I went off into laughing paroxysms and he got more annoyed. Eventually he fell upon me with kicks and blows and all the other servicemen who had missed the bus that night had to pull us apart.

On the way to camp next morning we called into a café and had fried egg sandwiches and Angus Duncan apologised. As he did so, I noticed his eyes on the waitress. They were full of longing.

He took Meta home a few more times after that, but the bedroom remained the route to the bathroom and no more. At the halfway mark on the course Angus Duncan failed in navigation, and a failure in that area couldn't be fixed by bribery, so he was posted to Toronto to start training as a bomb aimer.

I was happy enough to stay Gladys's boyfriend on a strictly petting-only basis. I liked the idea of having her on my arm, especially when she came to the airfield to see me play ice hockey for the station team. At the same time, the age difference niggled at me too. I mentioned it indirectly to Angus Duncan in a letter written just before I was due to take my final exams as an observer. 'I don't know whether to make any arrangement for after the war or not,' I wrote. 'I don't seem to be able to work out the extent of my liking for her. But I will certainly miss her.'

He rang me up to reply: 'Don't be daft, Paddy.'

'Why?'

'She's married to a major. He's in England. Meta told me.'

There wasn't a word wasted with Angus Duncan.

I couldn't understand why I'd never worked it out before. She'd killed two birds with the one stone, staying faithful to the major, and getting her nights out, plus squeezes on the sofa at the same time. My inexperience had guaranteed that there would be no pressures on her to open the hangar doors, as we used to say.

On our last night together she ran her finger along the outline of my flying badge, the 'O' with the wing, and touched the brand new stripes.

'Did you love me?'

'Yes.'

'Will you think of me, in England?'

'Yes.'

It satisfied her. She even cried that night. I never did tell her that I knew she was married.

As a romance it wasn't really worth a damn. I only hoped that she hadn't been laughing at me all the time. Certainly she'd listened sympathetically when I'd told her how much it meant to me to be an observer. She'd seemed to understand when I'd explained that graduation at Mount Hope was more than just a badge and stripes. It was a move all the way from a kitchen house in Belfast towards the like of Alec McCleod's detached, comfortable home on the Isle of Skye: from public elementary school to Edinburgh University. There didn't seem any point in letting her know that I knew about the major. When I was climbing the perimeter wire fence for the last time, I was thinking that her age could have been a bit of a problem anyway, if we'd ever become real lovers. Months later, when I was in England, I got a letter from Angus Duncan. It was to tell me that he'd married the waitress who'd served us with the fried egg sandwiches after we'd been out all night. I remember hoping it wasn't a rebound.

I was sent to Prince Edward Island next. This was where we were to be trained in ship recognition and ocean flying. Luckily neither subject gave me any bother. I'd spent most of my early days hanging around the Belfast waterfront. I could assess the tonnage and type of a merchant ship at a glance, and I could never understand how anybody could get it wrong. In action, submariners on both sides made huge errors in ship recognition, even mistaking small coastal packets for passenger liners. Coastal Command aircrew were far worse. The Navy used to split its sides laughing at some of the convoy reports transmitted by shadowing aircraft. Yet the intelligence rooms on operational airfields were practically wallpapered with silhouettes of ships of the opposing navies. It was only a matter of memorising them. By the time I left Charlottetown on Prince Edward Island I could take a fair shot at identifying the main classes in the British, Italian, US and German navies. As for ocean flying, provided the pilot knew his stuff and could fly a steady course, the very waves themselves kept the navigator right. The wind lanes from low and medium height lay obligingly along the lines of the drift recorder sight, giving wind-speed and direction as clearly as if the figures were inscribed on the face of the ocean.

We flew search patterns of all shapes and sizes: square searches; parallel searches; creeping-line-ahead patrols; patrols as from a moving base; radii of action; interceptions of various sorts; and by the time I had finished I was a true Coastal Command observer – much more at home over sea than over land. Over the land there was the continuing need for mapreading, interpreting the confusion of road, rail and river, all constantly shifting and changing under the aircraft. With ocean flying, the pencil and the chart, the sextant, the astro compass, the trailing aerial and the loop bearing made dead reckoning a dawdle. Prince Edward Island was for me a most interesting and enjoyable diversion. The sergeants' pay with its flying allowance was an added bonus. I'd have done it for nothing.

Then, as today, Canada was criss-crossed by signals from small commercial radio stations. Far out to sea, when the time came to turn for home, the lazy way to get a course was to tune in to the Charlottetown station and home onto it. But it was even nicer to work the course out first, then switch the radio on to check it, and finally cruise home to the music of Glenn Miller or Sinatra, warm in the old Avro Anson, the spring sun setting astern, landfall coming dead on the button.

In the evenings, walking in the snow, I could read the clear skies like a book. I knew Pollux and Betelgeuse, Rigel and Sirius; Alnitak and Alnilam were my very good friends. Bellatrix, Regulus and Polaris kept me company, and the sun and moon together in the early evening sky were there to fix my position simultaneously. If ever I needed them, I only had to reach for the sextant. That was when I discovered that the Father, Son and Holy Spirit were mathematicians. I was in communion with God when I determined sub-stellar points. I was at one with the Father, and he taught me how to pray when he taught me spherical trigonometry.

There was an officer pupil called Philip Leach on our course. He had the quickest mind I've ever encountered. In the classroom he was onto the point at issue before I had sorted the instructor's words into proper order. He flew with me sometimes, for we were required to fly in pairs, and never once did Philip Leach's bearing lines on the navigation chart zip through the dead reckoning position as mine did. He used to crib them from me, did Philip Leach, lightning brain or no lightning brain.

After graduation at Mount Hope six officers were chosen. They were selected by taking the marks gained from all subjects, including air work, and giving commissions to the top six. The total marks possible in this way were 1,200. The authorities then added a discretionary score, for what they called 'character and leadership'. So there was a maximum 1,200 marks for actual performance, plus a possible 2,000 marks for character and leadership. Trust the British: in this way they could have awarded King Kong a commission, but it didn't bother me any. At Prince Edward Island I simply didn't seem to be able to get over the fact that I had done it. I was a sergeant observer, and a frigging good one.

In the bog, I used to sit admiring the badge on my battledress blouse hanging on the hook above my head. I would sit thinking of St Athan, and the day down at the flights when I'd followed McCleod's eyes to the flier standing with the helmet in his hand, wearing the fur-lined flying boots. We were all aircrew now, McCleod, Burgess and I. When the flight mechanic saluted at the wingtip, as the aircraft taxied to take off, we were the sort of men being saluted. It was a lovely feeling.

Something else lovely was a trip to New York with Henry, a local restaurateur. His real name was Henryk, from the Polish, but he liked to crack on that he was English stock. I met him in a Charlottetown bar and he took a liking to me. He called the twice-yearly New York trip his lost weekend, and he wasn't far out. He was a good-looking man, about forty-five, with blonde hair beginning to grey, and features made the more handsome by Slavic cheekbones and tanned skin.

'Look, Paddy,' he said. For some reason nobody in Canada ever wanted to call me Sam. 'Look, Paddy, I'm taking you to the joints I frequent. We'll have one day and one night in New York. There won't be time to take you around anywhere else, so if you don't like my joints, tough goddam shit. When I get here I just let it rip. You go along, or do whatever you damn well like. Here's some dough,' and he handed me a roll of dollar bills that would have choked a crocodile.

'I want to see Jack Dempsey's,' I said firmly, 'and I want to see Radio City.'

'Yeah, I guess that can be arranged, but that's all, kid.'

Jack Dempsey's famous Broadway restaurant was crowded and, to my great pleasure, the champ was there himself, signing autographs, posing for photographs, and chatting. I was sitting at a table with a Scotsman who was training in Ontario, when this Yank navyman began to make fun of our accents.

'Oh, cheerio, old bean,' he mimicked, 'and say, guess who's a tooty fruit, then.'

His companions, civilian relatives by the looks of them, were embarrassed.

He was big, this sailor, but he wasn't unmanageable. I said nothing, but worked my nerves up to prepare for a duffy. The Scot beside me was all for heading out. People at adjoining tables were watching.

'Say, when you gonna let me see your panties, Nora?'

You couldn't come from Belfast and take much of that. I leaned over. 'Up yours from here to Detroit,' I said, in my friendliest manner. 'Your face and my arse could be twins, only your breath's the stinkingest.'

The sailor jumped to his feet, his friends held him back, I took hold of a handy stone flower vase in case he turned out to be good with the hands, but suddenly everything went quiet. We looked over to the one side. Jack Dempsey, the Manassa Mauler, the greatest fighter of all time, was looking in our direction. There appeared to be a slight frown on his face. This was in all likelihood the same slight frown that Dempsey had worn just before he'd pulverised Luis Firpo, the Wild Bull of the Pampas, in 1923. With one accord our two parties left the restaurant, patting the champ on the back as we went out.

Once on the pavement, the American sailor tore into me. A wild right caught me bang on the eye. I sank my own right a good six inches into his gut, below the belt, and was about to bend the vase over his head when the civilians and the Scotsman separated us, and the sailor left, calling insults over his shoulder.

'I leave you for one hour and this is what happens,' Henry sighed, as we drove to Radio City. I was lucky enough to catch one of the shows there, and then we went to a bar somewhere downtown, where Henry was greeted like Robinson Crusoe coming home.

The place was popular with lady factory workers. I got myself mixed up with some who were on what they called the 'swing shift'. This allowed them to booze up in the bar before going to work. Or was it after? I can't remember, but whichever side of the swing shift it was that the drinking went on, I certainly gave them a hand with it.

One woman in particular, called Kay, took a great shine to me. She insisted on taking me home at three in the morning.

'What about the swing shift?' I asked.

'That's just what you and me's going to start on, lover,' she said. The customers cheered as we left.

'I'll pick you up here at noon,' Henry called. He was sitting with a large brunette on his lap, buying drink the way they used to do it in the Klondike.

Her home was very nice. We had beer and cold chicken from the icebox, just as they did in the pictures. I said so to Kay.

'You and me, Paddy,' she said, 'we're gonna do a lotta things they *don't* do in the movies, too.'

And we did. By her bed there was this snapshot of an American sailor, in his dinky little hat. 'I don't know what he is to you,' I said to Kay, 'but is it all right if I turn his face to the wall? It was a sailor who gave me this black eye, after all.'

'Listen,' she said, 'that bastard's in Britain right now, and I'll bet he's double-crossing me. I want him to see me getting my own back.'

And he did, too. I had so many bruises from bites when I returned to Charlottetown with Henry next day that the black eye was just an also-ran.

We went back to Liverpool in a hell-ship called the *Louis Pasteur*. The journey couldn't have been in greater contrast to the dream-trip in the *QE*. Going home, we were riddled with dysentery, hounded by officious Army officers, and fed swill. The trip took the best part of two weeks, and it seemed like two years. The *Queen Elizabeth* had seemed like a peacetime cruise, but going home all the dangers of the Atlantic crossing were in our minds as we lay in our hammocks, well below the water line, making our way through waters practically owned by the U-boat hunting packs, in the spring of 1943.

Liverpool looked weary as we passed through but, surprisingly, the men and women in the dockside streets, so like my own in Belfast, cheered when they saw the bus loads of freshly-trained fliers leaving the quays. This was when the war was on the turn. Stalingrad had been saved. Montgomery was in North Africa on the heels of Rommel. The bombing of the German cities had begun in earnest. Liverpool had had the works from the Luftwaffe less than two years earlier. Its people were glad to see us in such numbers.

The general idea at Squire's Gate, near Blackpool, was to fly in Avro Ansons alternately as wireless operators and as navigators, gradually reaching the point where one man did both duties in the air. It was a bit hectic, rushing from the navigation table to the radio, but I managed it OK and, whereas I'd passed out below average at Mount Hope in Ontario, I finished with an average mark after the statutory eight trips from Squire's Gate, a fact which delighted me no end. At last I was as good as the grammar-school boys.

Next came operational training at Catfoss, near Hull, and my first introduction to the dreaded Beaufighter aircraft. By then it had the reputation of being a killer away from the enemy, as well as over the target. Crashes in training were so frequent as to cause airmen's eyes to roll when the name was mentioned.

The designer of the Beaufighter multi-purpose aircraft, Captain Barnwell, believed that it was possible to make the kitchen table fly if enough power was applied to it. It is a fact that the Beaufighter was only kept airborne by the brute force of hefty power plants. Without engines, she fell like a brick. The models in which I did my flying were the Marks VI and X, driven by the Hercules VI and VII respectively. She was blunt-nosed, heavily armed – four 20-millimetre Hispano cannon and six Browning .303 machine guns, with a single Browning for the navigator – and protectively armoured around the pilot.

The trimmer was so powerful that its control over the elevators could not be overcome if the pilot found that he'd adjusted it wrongly on take-off. Pupils and instructors, and even highly-skilled ferry pilots, had been killed as a consequence. When the Beaufighter was being introduced as a night-fighter the Merlin-engined Mark II took dozens of lives on training units. I myself had seen one spin in on take-off from Cranfield in 1941, killing the crew instantly.

The forward view from the cockpit was so unrestricted as to generate awe in the new pilots. Landing was made very tricky thereby. If the approach was not properly performed and a big engine surge was needed there was no way to tell if the tail was in the up or down position. This made for many memorable landings in my case. But then I drew a pilot who never did work out how to land a Beaufighter, or get one properly off the ground either. And when he did manage to get it airborne he made nothing but a pig's arse out of flying it as well. The published view of a highly-respected ferry pilot was that he never felt relaxed in the Beau until he was at 2,000 feet and 200 knots. For my part, when the full extent of my first pilot's ineptitude at the controls dawned on me, I was never really easy in my mind till the aircraft was down and tied to the ground. William Price was the name of this pilot. He was a nice guy, abstemious and moderate in all things, but as a wartime flier he represented the poorest return on public spending since Mussolini bought the horse that was to carry him, triumphant, into Cairo.

The take-off procedure was emphasised time and again by the staff pilots at Catfoss. Upon releasing the brakes and letting the aircraft roll a few yards, the throttles were to be opened steadily to the cruising gate at plus-two boost, leading slightly with the starboard Hercules engine. This gave sufficient horsepower per engine to render any further advance of the throttles unnecessary until a third of the runway had been covered. If this was not followed a swing could develop, a swing that was a unique feature of the Beaufighter. It occurred when the machine was almost airborne, and it occurred once every five or six times that William Price flew the bloody thing.

I was twenty-two years of age when we became a crew. It is no accident or coincidence that it was at the age of twenty-two that the first grey hairs appeared on my head. And this was before I had even met the enemy. I offered Mark Kerr, my messmate, half of my pay for the rest of the war if he'd take my pilot for his, but the very idea sent him white in the face.

But I was able to make trouble enough for myself, never mind William Price. In a bar near Catfoss, halfway through the operational training programme, I was sitting, half drunk, in company with several flight sergeant course instructors, Mark Kerr, the Derryman, and a Yorkshire navigator called Tom Cooney. The instructors were on six months' rest, having completed tours of operations from airfields in north-east England like Thornaby-on-Tees and North Coates. They were doing their best to put the fear of Christ into us and, more than likely, out of themselves, for they were due to go back on operations shortly.

'I'm telling you, you don't know the half of it,' one instructor, a Scot from the Lowlands, said, his eyes rolling. 'Those bliddy Focke Wulfe 190s. Jesus!'

'Even if you get away from the fighters,' his mate, a Londoner put in, 'the flak has to be seen to be believed.' He swallowed most of a pint of wallop in one absent-minded swig. 'The Jerries send flak ships along with the convoys, and those gunners have eyes like frigging hawks. If you're in the first wave it's goodnight, nurse.'

Mark Kerr and Tom Cooney were looking thoughtful. I wasn't. I was rather annoyed. I had a six-figure service number. By the second half of 1943 this was a most prestigious possession. It meant that I had enlisted shortly after the outbreak of the war. I was one of a privileged few. In a service where casualties had carried on steadily since the beginning of hostilities, length of service carried considerable authority amongst rankers and NCOs. Here I was, able to give the two instructors a year of service at least, and there the pair of them sat, making me feel like a rookie, in front of two of my more junior colleagues at that. Something clearly had to be done. As usual, I arrived instinctively at a form of reply that involved attack.

'I think you two guys ought to catch yourselves on,' I said, pretending to be angry.

The instructors went at once onto the defensive. 'What do you mean, Paddy?'

Never mind that I'd never flown a minute's combat duty, long service was written all over me the way I wore my uniform, the battered, creased forage cap tucked into my shoulder strap, the buttons on my tunic so smooth from polishing that the crest had almost worn off them. I'd been around. I was an old-timer, at twenty-two.

'Well, you're supposed to be the instructors for these lads,' I said, carefully leaving myself outside the rookie status of Kerr and Cooney. 'Here they are, full of enthusiasm, keen to get at the Hun, ready to do what you've done, survive a tour of ops in Beaufighters.' The instructors were pleased. So was I. I had placed myself alongside a couple of combatants. My six-figure number was getting its just due. 'And what are you doing?' I went on. 'You're sending them to their squadrons full of fear and trepidation. What way's that to carry on? For Jaze sake, you might be flying alongside them soon. You might need their courage and skill to find the target and sink the bastard and save you the job of doing it yourselves. You two should be glorifying war, instead of leaving this pair here as if they've just had a kick in the goolies. Look at them.'

Mark Kerr, who loved to see me at this old soldier act, put on a look of abject fear, spilling his drink as he picked up his pint glass. Tom Cooney really did seem worried. The two instructors admitted at once that I was right, and rose to buy a round of beer. Delighted with myself, I got up and made my way in the direction of the gents. That's when I bumped into the air ace.

It was accidental. 'Whoops, old thing,' I said, full of beer and light repartee. He was turning away from his company at the bar. I had knocked his pink gin down the front of his tunic. It splashed the bottom of his left-hand pocket, just under the ribbons of the DSO and the DFC. He was a wing commander, fair-haired, thin-lipped, very good-looking. And very angry at me.

'You bloody fool,' he muttered, reaching for a hankie.

'Sorry, sir,' I said, not looking in the least sorry. I didn't like being called a bloody fool, so I gave him, eyeball to eyeball, a look that told him what might have befallen him if he hadn't carried the smoke on his shoulders, by way of wing commander's bars. The officer straightened up. His companions, a very attractive young lady and a man who was dressed in tweeds and looked like her father, watched with amused interest. The place was packed with sergeants and other ranks. I had to brace myself against the push of the crowd, and was having trouble in making progress, so I couldn't make the escape that the situation screamed for.

'Take that hat from your shoulder strap and put it on, Sergeant,' the wing commander snapped. 'And stand to attention.'

The familiar speeding-up took place inside me. I went right over the top. The guts tightened, the breath was sucked in, the eyes narrowed. I stood to exaggerated attention, quivering like a reed, my head so far back that I was looking at the ceiling, my eyes wider than saucers, and I wore a look of abject terror.

'Sir,' I shouted, 'I'm at attention. I'm at attention like anything. Atten . . . shun!' All around, the drinkers fell silent. 'Ready for further orders, sir,' I yelled.

The girl burst out laughing, and I gave her a wink. The wing commander was furious.

'What's your number, rank and name?' he snarled. At the top of my voice, I rhymed them off.

'That's telling him, Paddy,' a voice shouted from the back of the crowd.

'Report to me in the morning, Sergeant,' the officer said. I nearly tore my vocal chords apart, acknowledging the order. The wing commander turned his back to me, and I carried on to the gents.

The two instructors were awestruck when I got back. 'That's a pig, that guy,' one said. 'Christ help you in the morning.'

When my eyes were closing in the hut that night, I cursed that short fuse of mine, and wondered what would happen in the morning.

There was no doubt that he had carried his umbrage through the night. Standing in front of his desk at Catfoss next morning, I listened while he told me that he was considering putting forward a recommendation that I should be taken off aircrew duties and reduced to the ranks. 'What do you have to say to that?' he rapped.

'I respectfully submit that what I did last night doesn't merit that sort of punishment,' I said, in as reasonable a voice as I could manage. I wasn't feeling in the least cocky. Indeed, I was a mass of nerves. 'I am of course sorry for last night, sir,' I went on. 'I had too many drinks.'

The officer sat forward. 'I don't like you, Sergeant,' he said, 'and I don't like your attitude.'

He relaxed. 'And talking about your bloody self-confident manner, you're only a bloody pupil here. You've done nothing in the Air Force. Nothing.' He waited for a reaction.

His gallantry ribbons seemed to have floated off his chest and become enlarged until they formed the whole backdrop to the scene. I seemed to see his DSO and DFC ribbons covering the entire wall behind him. There was no doubt at all in my mind that this man was flaunting them. It was the only time in the entire war that I ever saw it happen. Once again I felt my control begin to slip.

'I know I'm nothing,' I said, 'but at least I'm game for a go in this war. I can't do any more than that.'

'Look, you,' the wing commander sneered, 'the donkey-work has already been done in this war. People like you don't matter a damn.'

Now it so happened that, only two days before, I had been to Waddington, a bomber station not very far away, to look up an old friend, a bomb aimer on Lancasters. There had been a raid on Berlin, one of several carried out against the same target in recent times. Losses had been heavy. I had been present in the late afternoon, when the crews who had taken part in the previous night's operation were appearing, after their rest, back in the sergeants' mess. I had been quite surprised at the effect on some of the aircrews of night bombing over the Reich. Several of the sergeants seemed, to my eyes, to be unfit to fly, their nerves were so badly gone. My friend had completed about twenty trips and was drinking for effect, rather than for social reasons. The times were bad for Bomber Command. The German night-fighter system had been refined to the point where high interception rates were the rule. Skill was no longer any guarantee against being shot down. At that stage of the war, it was only raw chance that dictated whether or not the bomber would return to England after its raid. The Ju 88 night-fighter came on its target from astern, and from underneath. No matter how vigilant the rear gunner on the Lancaster or Stirling bomber, the odds were on the fighter coming close without being seen. And once that happened there was little hope for the RAF crew. Twenty-millimetre cannon made an abattoir of an aircraft's interior, and the Germans usually fired from a hundred yards or thereabouts. Although Beaufighter operations were, statistically, the most dangerous being flown, I was glad that I wasn't on the heavy bombers. To keep company each day with such vast numbers of crewmen, to make friends with them, and to find so many missing so often, made it a hell of a nasty war to fight. At least when a Beau crew went into the sea, only two men went missing. On Lancasters, seven men disappeared for each Ju 88 kill.

'You don't matter a damn,' the wing commander said again. By then I didn't give a shit.

'I wonder what would happen to you if you were to go down the road to any of the bomber stations and say that,' I said, my cheeks blazing.

In the corner of the room a flying officer, an elderly man with First War ribbons, had been observing, in silence, probably as part of the disciplinary proceedings. He suddenly cleared his throat. I caught his glance. He frowned slightly, gave a tiny shake of the head. But it was too late.

'What the hell do you mean?' The wing commander sat up.

'What I mean is this,' I said. 'Bomber Command lost over seventy kites a couple of nights ago. Go to Waddington or Leeming and tell them that you've done the donkey-work in this war. Let me know what they've to say to your theory.'

The older officer rose, bent over the senior man, whispered something, then he went to a window and stared out.

'Get out, you insolent bastard,' the wing commander's voice was hoarse.

I saluted and left. I never saw him again.

With William Price I flew in the Beaufighter over the North Sea on air firing and navigation exercises. We threw the aircraft all over the sky in mock attacks by Spitfires, and I found that I liked the Beaufighter work, as far as my own side of it went. I had no problems at all either with radio operating or navigating. I was graded above average at the end of the operational training course, but still I was disappointed, because William Price was passed out as fit too, despite a whole range of flying eccentricities. Others had noticed them too. It got to the stage where none of my friends would lend me money.

I went with William Price to Filton, the airfield beside the Bristol Aircraft Works, to collect a brand new Beaufighter and fly it to Islay for proving flights. While I was there I slept overnight in a hut with only one other occupant. He was a flight sergeant fitter, about fifty years of age. At breakfast he was bright and cheerful, full of reminiscences about his years overseas as a regular airman. But in the afternoon he sat on his bed, silent. I paid no great attention since I was composing a letter to a woman I'd met in the Mauretania bar in Bristol the previous night. I was telling her that I would probably be killed in the war, and that it would be some consolation to die knowing that I had been granted the greatest thing that a woman can give.

An odd sound made me lift my head. It was the old-timer. He was sobbing, holding a letter. I went and sat on the empty bed near him: 'What is it?'

He handed the letter to me. It was what we in the Air Force called a mespot. In the American forces it was known as a Dear John. His wife had met someone else. A factory worker. She was already living with him in Yorkshire. The letter ran to only a dozen lines.

'Jesus Christ,' the other gasped, 'she's forty-bloody-seven.'

His hair was iron-grey. He had false teeth that fitted badly, as my father's had done. They made his upper lip jut out, just like his. He was holding his forage cap, rolled up, and every so often wiping his eyes with it. His eyes were red-raw.

I looked again at the letter. The pen was still in my hand. I wrote at the bottom of the cheap, lined paper 'PTO'. On the other side I wrote: 'Tom read this for the first time in a dirty, draughty, nearly-empty hut, located in a dismal, temporary airfield dispersal. He can't help being apart from you. He is crying his eyes out as I am writing this. I think you're a bit of a shit. Sam McAughtry.'

I addressed an envelope, used one of my stamps, tucked the Dear

John letter inside, and posted it in the sergeants' mess postbox. With it I sent also the letter to the woman from the Mauretania bar. I never saw the flight sergeant again. Unexpectedly, we took off for Islay next day. My ninety-seven per cent failure rate with women was holding up. Now I would never know about the woman in Bristol.

Five hundred miles out into the Atlantic. The height is 1,000 feet. The sea is as grey as the grey of an old seadog's cheek. The sullen waves are topped with suds. The water is the colour of the water wrung out of the floorcloth that washed the floor of 118 Hillman Street, over there to the west. Hillman Street is over the horizon whence come the radio bearings, the back bearings that must be the same value as the back track, otherwise we're lost.

Mart, my brother, was lost in the Atlantic. The suds that dive off the wavetops can be made to perform work. They can be made to lie along the grid of the drift recorder, and the angle that they make with the lubber line of the aeroplane is the difference between the heading and the track made good. If we were to fly over a sailor floating dead in the Atlantic, we could make his body perform work by having it lie along the grid of the drift recorder, and the dead man would signal the value of the angle between lubber line and track made good.

The sky is dove-grey, and it is leaden-grey, and it is greenish-white. It has high, piled-up, tightly-pinned hair, or it is loose, tumbled, its ends flying away. A cloud opens suddenly to let a slanting ray of brighter light through from the high cirrus, and the mares' tails, and the blue reaches that mark a place where we never go, in our Beaufighter.

The base of the cloud is the point on the graph where the environmental lapse rate and the dry adiabatic lapse rate meet. And Buys-Ballot's Law states that an observer in the northern hemisphere with his back to the prevailing wind will find the area of lowest pressure on his left-hand side. Knowing these things is supposed to make the Atlantic sky less intimidating when you're at 1,000 feet in a Beaufighter 500 miles from the nearest land. But my father and his father and his before that at the very least were seafarers, and with that sort of breeding there's no intimidation to start with. I never did need lapse rates or laws to go out over the ocean with confidence.

In all the life that may lie before me there will never be any pride that will touch the pride that I feel in being a good navigator. The college boys will return from this and forget about it. It will stay with me forever.

'Alter course one-eighty. Steer zero-eight-three compass for Islay.' William Price throws her about, and performs work for me . . .

Islay sits only a few miles from the North Irish coast. It doesn't take the day to be very clear to see Antrim's cliffs and Rathlin. The island's western coastal waters are littered with wrecks. The skeletons of many of these are to be seen here and there buried to the gunwales in sand. And during the war Islay bade fair to lay up as big a collection of wrecked aeroplanes. Certainly, during my time there, the Beaufighter casualties were hair-raising. The winter of 1943 was a bad one on Islay. The weather had a habit of closing in without warning, but we were required nevertheless, on the Beaufighter ferry unit, to fly our aircraft in conditions which would have caused an operational squadron to stay on the ground.

The people were unfriendly. The island was a gloomy burial mound of a place, a hungry hole. On the Beaufighter unit we stayed mainly in camp, the pubs were so lacking in warmth. I made a sort of partial inroad into the islanders' community, when I came upon a man in a pub in Bowmore playing the bagpipes very badly. It so happened that I was able to play the things and, watched closely by a group of nine or so locals, I tuned the instrument and then blew a selection of marches, strathspeys and reels. When I had finished, the man who'd been murdering the pipes earlier grunted.

'Aye, ye've seen them things afore the day,' he said. It was the only phrase spoken by an islander to me during my stay that could be said to lack hostility.

From here we flew our aircraft out far over the Atlantic, to extreme limits, testing them for battleworthiness. A few days after we arrived three Beaus came over the airfield at Port Ellen, back from a long haul over the sea. One crashed by the side of the runway and the other two flew into some hills a couple of miles downwind. All three crews were lost. Only a few weeks earlier the six fliers had been posing with me for a group photograph outside the hangar at Catfoss.

William Price still swung on take-off and landed too high up. They were faults he seemed unable to iron out. He was an officer, and officers literally got away with manslaughter, as far as airmanship was concerned. Some of the sloppiest pilots I met in the war were commissioned at flying school, and some of the best were commissioned in the field.

Back in Belfast on my last leave an old lady called Oakman, who lived opposite us, had called me across the street. 'Here's a lucky charm for you, son,' she said. 'It saved my Willie's life in the first war, so maybe it'll help you in this one.' And she handed me a silver threepenny bit on a chain. I had worn the lucky charm through the trip to Islay from Filton, and on two early flights out over the ocean. Then, on his third trip in the new aircraft, William Price did it properly. He swung so badly on take-off that he went right off the runway at ninety knots, ran up a huge heap of stones and spoil, and tore the Beaufighter almost in two, setting fire to the port engine. There was nothing but open space astern of my seat, so I dropped out to the ground, somewhat bruised and shocked. The station ambulance drew up and I was taken aboard, but just before the ambulance door closed I remembered Mrs Oakman's lucky charm. Tearing it from around my neck, I threw it as far across the grass as it would go.

It was only an hour later that it dawned on me that the thing had, in fact, brought me luck. Not many watchers on the airfield had expected William Price or me to survive the crash. I was grateful in the end for the old lady's luck token, but reckoned that it was a whole lot too effective in the case of William Price.

6

Into action

Flying down from Islay to Portreath airfield in Cornwall, I could hear on my headset the Portreath Beaufighter navigators chattering in Morse to base. There were operational Beaus based there and these ones were returning after a sweep of the Bay of Biscay. These sweeps had much of the flavour of the first war about them in that they were often provocative in nature, and many times attracted the twin-engined Ju 88s, the Beaufighter's natural enemy, to combat. Listening to the laboured Morse, the thing that impressed me most was the use of the operational prefix 'O' to their messages. This meant that they were real fliers, not beginners, like us. After landing, I couldn't wait to run the squadron to earth and talk to the navigators.

We were all known to each other, the Beaufighter boys. We were a tight club, whose members had a habit of leaving suddenly, without formal resignation. The monthly magazine *Flight* lay around every RAF mess. It carried details of all aircrew casualties, and we saw on the casualty lists, every month that passed, the names of men who had trained before us, on this station or that. The medal lists, oddly enough, were not as well sprinkled with their names. Spitfire photo-reconnaissance boys got a medal when their tours ended, although they lost few machines. Night-fighter crews were decorated after only one victory, yet their job was one of the safest in the Air Force on the operational side. But the Beaufighter ship-busters and before them the Beauforts and Blenheims, and the Fleet Air Arm Swordfish and Barracuda crews, went in during daylight against horrific flak defences with little in the way of medal ribbons to show for it. The worst of it was over by the time I got into things, but what was to come was a long way from being a joke. I picked the brains of the Biscay squadron, and stood by to fly the long leg out to Rabat in North Africa, shutting my mind off to the fact that it was William Price who was to take me there.

The thing was, we had to wait for a following wind before taking off. It was a six-hour trip to Rabat, just about the maximum for the Beau, and a head wind would have brought us down in the Bay, or in Spain or Portugal. The wind wasn't right when I got to Portreath, and we were put on full-time standby until it was. It was hammered home to us that, when the call came for take-off, it would come at some ungodly hour of the morning, and that we should retire early against this contingency. I spoke to the meteorology officer.

'What would you make the odds against take-off in the morning?' I asked.

'Two to one,' was the reply.

They were good odds. I went out to a little town called Redruth and got off my mark with a dark-haired number whose name was Sarah. I told her that I had been shot down three times and that once a German rescue plane had landed in the Bay and tried to take me on board.

'Unfortunately for the two-man crew,' I said, 'I happen to carry a Derringer under my jacket. I shot the crewman and the pilot scrambled back into the seaplane and took off.'

'How did you get back?' she wanted to know.

'The German flew over Portreath and dropped a message,' I told her. 'The message, with two red ribbons tied to it, said: "The sergeant with the north Belfast accent is in position so and so. I salute him. He is the bravest foe I have met. He is also the most cunning. Who would ever have thought of hiding a Derringer? We Germans admire such ingenuity. I happen to own a big factory in the Ruhr. After zis war I vill give zis Belfastman a job at a very high salary. Give him my best regards. His foe right now, but not for ever, Alec Richthofen. PS, I too am against the Pope."' I drew her close. 'But let's not talk of such things now,' I said, 'let's talk about that split skirt you've on. Your legs are frigging brilliant, did you ever know that?'

Next night I decided to take another dive into Redruth. I had the woman on the skids. One more push and the job was a good one.

'I've been writing a story especially for you,' I whispered to Sarah in her front sitting room with her mother hovering in the hall outside.

'Are you a writer, then?' she asked.

I nodded, modestly. 'Only four novels accepted so far, plus six plays and an appreciation of Thomas Hardy, but after the war, if I recover from the battle nerves, I hope to live by my pen.'

'What's this story?'

'It's called "Cupid and Sarah",' I said. 'I spent the whole day writing it.' Then I swung into the story of Cupid and Psyche, from Apuleius's *The Golden Ass*. All through, I substituted the name of Sarah for that of Psyche. '. . . So beautiful Sarah finally became a goddess much to the chagrin of her sisters,' I finished.

'Oh, what a beautiful story,' she breathed. 'Ar,' she added, in her Cornish way, 'but oo are talented, aren't oo.'

'You kind of inspired me,' I told her. 'The only time I ever felt this way about any girl was when I was in love with one called Aggie. Aggie was blown to smithereens in an air raid,' I said. 'We were lovers, and I've missed the physical side something desperate ever since.'

'What was her story about?' Sarah asked.

'It will never be told,' I said, shaking my head, 'but it was called "Aggie Memnon". That's all I can say at this time.'

Just with that her old lady knocked, wanting to know did I intend to sleep in her front sitting room all night.

'See me here tomorrow night,' Sarah whispered. 'She's going to the Mothers' Union.'

'Am I under consideration for the physical side of things?' I asked. There was the briefest of shy nods in reply.

I walked it the four miles back to Portreath on a cloud, got into bed at a quarter to three in the morning, and was called at three o'clock. We were on our way to North Africa. I was furious. I'd been so looking forward to telling my girl in Redruth this new story I'd written, called 'Sarah of Troy'.

The line of take-off at Portreath took the aircraft over the edge of a very high cliff. On the same day that I left there, a DC3 troop-carrier, with eight or nine of the lads who had trained with me in Canada on board, tipped over, one wing low, as the current of air from the cliff met it. The pilot couldn't recover in time and all aboard were lost. We weren't to know this, William Price, Demon

Pilot, and myself, as we waited our turn amongst over a hundred aircraft to take off. I was absolutely whacked after losing the whole night's sleep. I had been handed a thick sheaf of briefing material, but was totally unable to understand it. William Price told me off when I fell asleep in the back seat.

'You never see me going out and roistering,' he said.

'No,' I muttered to myself, 'you have to keep yourself in training in order to do your celebrated circus trick, landing an aeroplane on its arse, or taking off sideways.'

Right enough, I thought to myself, we were a right crew, the pair of us. I went through life either building a hangover or shaking one off, whilst at the same time I could do my flying job on my ear, whereas William Price treated his body like a temple, abused not one corpuscle of it, but tried to kill the pair of us every time we went into the air. I thought that I deserved a good pilot, while he thought that he deserved a sober navigator. It never occurred to me to complain to the high-ups about his airmanship. In a way, I regarded William Price as divine retribution for my sins, and in any case I was sure that he'd be protected, since he was an officer, after all.

We took off at 0930 hours after sitting in the aircraft for five hours. We weren't on speaking terms for the first half of the trip, except when I spotted a Ju 88 in the distance and smartly gave a course away from it. William Price, during this first, albeit long-range, encounter with the enemy, showed all the fortitude of a busted jellyfish. I found myself, after we lost the German, singing that song from the film *Dawn Patrol*, the one that goes, 'Here's luck to the next man to die', and my pilot threatened to put me on a charge. Then he started to express doubts about my navigation, when our Cape Finisterre landfall looked like being late. But when we hit the Cape and began to hop at low level down the Portuguese coast, we were friends again.

On the way over to Rabat I looked up a few frequencies and got to chatting with a French ground station in the heart of the Sahara. Our exchanges were in the international Q code, but when he signed off he wished me 'Good luck, old man'. It was a tribute to the fighting man. I was highly delighted with myself.

On 7 December 1943 we took off from Rabat to fly to Castelbenito, on the way to Cairo. We swung sixty degrees on take-off and William Price was severely told off by the tower. After only two hours' flying, my pilot told me that the heat was making him feel unwell, and could he have a course for the nearest airfield, please.

I gave him a course for Biskra, in Algeria. In landing there, William Price broke an aeroplane's back once again. Our kite was a write-off.

Having hitched a ride in a Dakota to Almaza transit camp outside Cairo, I met up again with Mark Kerr, who had trained with me at Cranwell and in Canada. With him was another Derryman, Mick McDermott. Both had been educated at Foyle College, Derry. Mick had been head boy there. It was through Mark's pilot, Len, that I first saw the British colonial way with the natives.

We'd just had a meal that tasted wonderful to us, shrivelled as our stomachs were from tight wartime rationing. We were sitting back, smoking, over coffee. Behind us a waiter, a handsome man in snow-white robe and red fez, respectfully bowed and left the bill on the table. Without even breaking his discourse Len lifted the bill and threw it onto the floor. We three Irish were horrified. I looked for signs that it was a joke, but Len was, in fact, quite peeved.

'Cheeky bastards,' he growled. 'The time to give us the bill is when we're leaving.'

Len had been overseas for two years. I remember hoping that it wouldn't do that to me. The waiter picked up the piece of paper and bowed again, apologetically.

Later, at Buq Buq, I was to see a British Army corporal bartender in an officers' bar tell a visiting Egyptian officer who asked for a beer to piss off, no wogs were served.

We queued for Christmas dinner at Almaza. I was first. Each item of food was served by a smiling Egyptian. After I'd collected my chicken, potatoes and peas my plate was seized by a native and, before I could stop him, he poured custard all over my dinner.

'You stupid bastard!' I yelled.

'There, didn't I tell you you'd soon get wise to the wogs?' Len said.

I went to Shallufa to learn torpedo tactics, carrying out sham drops against the merchant shipping in Suez Bay. During a break I went to see what a brothel was like. The one chosen was an official British Army knocking shop for sergeants, in Rue des Souers, Alexandria. I was introduced to it by a ground-staff Welshman who visited the same girl there each month. Inside, it proved to be most businesslike, not the least as I'd expected. A military policeman entered customers' service numbers in a book. After the coupling the girl gave the customer a numbered disc, and this was entered against the man's name on the way out. Thus, the epidemiological trail was marked, in case disease was transmitted. It was one of the benefits of fighting for a thoroughly colonial power.

I wandered slowly along the corridor, sorry I had come, totally unable and unwilling to rise to the occasion, when suddenly a door opened and a coal-black momma appeared. She must have weighed fifteen stone. Her teeth contained enough fine metal to have shaken Johannesburg, had it been freed onto the market.

'Hello, soldier,' she cried, clasping me tightly to a pair of jujubes that felt like zeppelins, ends-on. My face buried in flesh as deep and slippery as buttered octopus, the negress carried me into her room, my feet six inches off the floor.

Inside, holding me firmly with her left hand, she went through my shirt pockets with the right, until my money was revealed. Peeling off the required amount in piastres, she let me go and dropped onto the bed, hauling her shift up to her chin. With her right hand she reached under the bed and found a chamber pot. This she placed in the centre of the floor, then, with a wild contortion of her features obviously meant to be coquettish, she arched her back. The overall effect was reminiscent of a slag heap, covered in black cellophane, undergoing subterranean slippage of some violent kind. I suddenly came to my senses and hit the trail for the front door. This one made the mares of Diomedes, who lived on human flesh, seem like Dartmoor ponies.

The MP at the door stopped me: 'What number?'

I told him that I hadn't any.

'Oh,' he sneered, 'a poofter, eh?'

'It was for your sake I left,' I said. 'That black giantess in there's almost certainly your mother.' Bought sex might have scared me, but policemen never did.

At Shallufa, I woke up one morning to find one of the native sanitary men going through the pockets of my battle-dress jacket, which was hanging on a tentpole. There was a powerful lot of thieving going on at the time. I sat upright in bed and the Egyptian fled, but I had seen his face. After breakfast I went to the guardroom and reported it. It was a decision I came to regret. In mid-morning I was called to the guardroom. Lined up inside was a row of unsavoury natives, sanitary men and general rubbish removers. I was taken along the line and had no trouble in identifying the culprit.

A corporal of some Palestinian force or other, an English-speaking foreigner, dragged the wretch out of the line and belted him across the mouth with the back of his hand, drawing a fair amount of blood. It was like hitting a woman. The guy had no idea how to use his hands, not that it would have done him any good. He steadfastly refused to admit that he was the one who had been searching my blouse – nothing was missing from it – and the more he denied it the more often he was beaten, until eventually, on his knees and streaming with blood, he confessed to a number of pilferings.

'What'll happen to him now?' I asked, as he was taken away.

'Oh, he'll go to jail in Suez for six months and his daughters'll go on the game,' I was told by the corporal, whose knuckles were cut and bleeding.

'God almighty,' I breathed, 'if I'd known that I'd have given him my dough,' I said.

'You stick to your job, Sergeant,' the other said, 'and let me do mine.'

'Too true,' says I, 'and I hope you keep enjoying it, the way you did just now.'

At operational training unit and with the Biscay boys at Portreath, I'd questioned and cross-questioned the best fliers regarding operational flying. I was insatiably curious to know all the ways in which my life could be lengthened as I did my bit for King and country. Luck didn't interest me. I didn't carry little teddy bears or rabbit paws like so many other fliers. Luck could keep me up or bring me down, I had no control over it, so there was no sense bothering about it.

I wanted to know how to get over the first few operational flights when, statistically the chances of destruction were highest. And again and again the same lesson came through: the best way to lower the odds against being shot down by flak from ship or shore installation was to go straight for the target, break as low as possible, and do the wobbling on the way out. As far as single-engined fighters were concerned it was a matter of hitting the trail – if there was one. If there wasn't the scope to get away, then the thing to do was to turn and have a go. The Beaufighter carried heavier armament than any German or Italian fighter and the enemy knew this as well as anybody else. If a dogfight developed the Beau's number was up, but sometimes that initial show of belligerency was enough to open a route to safety.

William Price and I learned more of this sort of information on our way up the desert to join our chosen squadron at Mersa Matruh. Sitting in the back of a bouncy Dodge fifteen hundredweight was a flight sergeant pilot who was halfway through his tour of operations on the same squadron. He'd seen action during the evacuation of Royal Marines from the islands of Kos and Leros a few months earlier, but I had to dig the facts out of him. He sat, huddled, with his back against the cab, staring out across the drab desert, his red, burned face set in an expression part melancholic, part frightened.

'How do you feel, with twenty-one trips done?' I asked him.

'Nothing,' he said. 'And for Christ's sake shut up. You'll know soon enough what ops are like.'

Later, in the sergeants' mess, I found that the same pilot was very fond of the beer. He loosened up after I'd joined him in a jorum or two.

'I don't much fancy the look of that pilot of yours, Paddy,' he said. 'He seems a nervous bastard. You should have seen him when you were asking all those questions on the way up.'

'Piss off, you,' I said. Right then I hated him for reminding me. On the road up, William Price had been chewing his knuckles like they were pig's feet.

On our first day at the flight dispersal at the landing ground south of Mersa Matruh, the coastal town which had been the scene of so much fighting in the desert battles of over a year earlier, one or two of the old sweats among the crews were trying to put the wind up us. As a polished line-shooter myself I wasn't affected by the stories of slaughter and destruction that they told, but was merely admiring or criticising the way that they did it. Not so William Price, however. His thin, dark-shaven face twitched as the veteran crews described what it was like over enemy convoys and harbours in the Aegean Sea due north of us. William had already formed the opinion that I was a hazard because of a habit I had of not taking notes at briefings, and not carrying huge masses of maps and charts and frequencies whenever I flew, the way so many other navigators did. In the air he had already rebuked me many times for smoking in the back seat. All of a sudden his querulous Yorkshire voice would pipe up: 'Is that cigarette smoke I smell, Paddy?'

'Jaze, William, but you've got cigarette smoke on the bloody brain,' I would say, drawing happily on my fag. 'I hope it's not an engine or anything going on fire,' I would add. 'Keep you your eye on the cylinder-head temperatures. Sometimes oil smells like Woodbine, you know.'

I must admit though, I wasn't too happy with things at the start of my fighting days in February 1944.

The role of our squadron, and two other Beau outfits, together with two South African Baltimore light bomber squadrons and Wellington bombers for night torpedo work, was to interrupt supplies to the German garrisons on the Aegean islands, particularly those on Rhodes and Crete. These garrisons had been established in 1941 after the Germans had taken the area. The Greeks were suffering terribly under the occupying forces and, as well as that, the islands and their airfields offered a distinct threat to the Allies' Mediterranean communications. The German and Italian garrisons were supplied by sea, and by Ju 52 transport aircraft, flown in by night. There were German and Italian fighters on the bigger islands. It was a strange, isolated little war that we fought. The big war had passed us by long before. Montgomery, Mark Clark and Alexander were on the toe of Italy, many miles away. The famous desert forts were all around us, empty. The sad debris of battle was everywhere – rusting tanks and transport, crashed aircraft, grave mounds in the open desert, wooden posts and steel helmets at their heads.

I often wandered around the old fortifications, despite warnings of mines in their vicinity. There was for me a curious fascination in the ragged uniform jackets, green for Italy, grey for Germany, khaki for the Allies, lying in gun pits that were filling up again with sand. There were old letters from home, water bottles still holding the smell of wine, uniform field caps representing the armies of both sides. Not far from Gambut, near Tobruk, I saw a small anti-tank gun, carefully kept apart. It was smashed in on one side of its protective shield. A back-street lad from Gateshead in the north-east of England, Adam Wakenshaw, had won the Victoria Cross behind this gun, nineteen months earlier. With his left arm blown off he'd continued to fire until a shell blew him to pieces. Today the gun stands at Brancepeth Castle, near Durham. When I saw it, in the old silent battlefield, it was easy to picture Wakenshaw's sacrifice, and those of the Axis and Commonwealth troops around him.

There were the Bedouin Arabs forever mooching around the old forts too. They made their women walk ahead of them in case there were mines about. I sometimes went into Bedouin villages, for curiosity. The women I saw were always quiet and thoughtful. I remember thinking it was no frigging wonder.

In our private war we knew the names of all the bigger ships supplying the garrisons. They couldn't get out of the area. They were well and truly bottled up. And on the islands and the Greek mainland itself there were British SAS men operating. Sometimes they came up the desert and stayed with us, and a fierce bunch of headcases they were, given to knocking each other's faces in, when the drink took hold. Still, I was the last to complain about that. An incident in which I figured at Shallufa had followed me to Mersa Matruh and ensured my reputation as a wild Irishman for the rest of my tour.

On my first night at Shallufa I'd gone up to the bar and ordered a bottle of beer. It was refused. I pointed to a nearby table, where a party of sergeants could hardly be seen behind a wall of full bottles.

'They're permanent staff,' I was told, 'you're a pupil. You get none.'

I walked over to the table. 'Sell me one of your beers,' I asked.

'Piss off,' this little Glaswegian snarled.

What the drinkers and the bartender didn't know was that I had been out drinking brandy in Suez earlier. Brandy is a stimulant. Much stimulated by Jock's wretched manners, I took hold of the table and upended it, beer and all. Then I sent the Glaswegian flying the length of the mess and proceeded to sail into the rest of his party. That was all I remembered. I awoke next morning, stretched, thought to myself that my face felt very tight, got out of bed, looked in the mirror, saw nothing but weals and knuckle marks, and wondered why my hair should be sitting in little, tight curls. My tentmate, a dry-witted London Jew, soon told me.

'They had to put you under the shower last night,' he said. 'You went berserk. And even at that, you wouldn't go to bed till you were given a beer.'

Going into breakfast that morning, I decided to get things over with the minimum of delay. I never was one to sulk in my tent. I joined the permanent staff for their meal, sitting beside the little Scot.

'Morning, lads.' I said. 'I believe you already know me.'

Three of them had contusions. After a moment's silence, they moved over for me. After that, I got the same beer ration as the permanent staff.

When I first walked up to the bar of the squadron sergeants' mess in the desert and asked for a beer, the guy behind the bar said, 'Sure, Paddy. Here, take two, take the lot . . .'

After some local flying William Price and I went off on our first operation. It was a sweep of the central Aegean, looking for shipping. Two days previously six Beaufighters had done a similar sweep. They'd run into an escorted merchant vessel and attacked it, without sinking either target or escort. One Beaufighter was lost. The thing was in William Price's mind a great deal.

'Do we lose one every trip?' he'd asked the flight commander, anxiously.

'How the hell would I know?' was the reply. 'It's bad luck to count.'

We took off at 1400 hours and came back at 1930, and didn't

even see a rowing-boat. At that time there was nothing rushed about our flying. After one trip there was a nice little rest of four days or so before the next one. I joined the old sweats at the bar, feeling a hell of a fellow, with my one operation done.

'You don't feel so bad, with five other Beaus around, up there,' I said to one navigator, a nervous wreck called Alec Linfield.

'You didn't maybe,' he said, 'but the rest of us did all right. That bloody pilot of yours was wandering all over the sky, jinking and weaving and bobbing up and down like some sort of nut. What the hell's the matter with him?'

'He's afraid of being killed,' I said.

'Well, tell him from me that he'd better fly straight and level from now on,' Alec said, through gritted teeth, 'otherwise I'll shoot the bastard down myself.'

'If you do,' I replied, 'I'll never speak to you again, ever, ever.'

But William Price's nerves were no joke. Four days later we were sent up on a night reconnaissance around the islands, and I was kept busy finding my way at a fairly low level around islands that had hills looming everywhere. Over the harbour on Paros island I said to William Price, 'That starboard engine certainly sparks a lot, doesn't it?' A stream of red sparks had appeared outside my cupola, close enough, it seemed to light my fag.

'Maybe it's sand in the engine,' William Price suggested, comfortably. Just with that the sparks turned orange and blue as well as red, and began to grow to enormous size, shooting not only past us, but up from under. It was like the Hallowe'en fireworks, multiplied by about fifty. Red onions and turnips were coming at us from all directions. Loud explosions were occurring at or near us.

'Hit the bloody trail, mate,' I shouted, 'we're getting the shit knocked out of us.'

It was the first time we'd seen flak, and it was nearly the last. When we got back our aircraft was riddled like a sieve. There were holes all around my part of the aeroplane. Price had taken the long way round, avoiding the ground fire.

'I can't stand much more of this,' William Price said to me under his breath, after debriefing.

'You can't stand much more of it?' I said. 'Sure it was nowhere near you. The stuff was parting my hair. And another thing,' I went on, 'when I told you to get offside away from that harbour, what the hell made you hang about? Are you getting keen all of a sudden?'

'I didn't hang about,' William Price said. 'I got out of there as soon as I could. I suppose my reactions were slowed by the flak. I turned right instead of left.'

'Flak's supposed to speed you up,' I came back, a note of desperation in my voice. 'What the hell are you going to do against the Messerschmitts? Exchange visiting cards?'

Next day the CO, a nice, easy-going chap name of Woodruffe, sent William Price up to do some flying exercises. I made an excuse and watched from the ground.

'Why don't you fly with your pilot?' I was asked by my flight commander.

'Does Joe Louis get into the ring with heavyweight boxers when he's not being paid for it?' I replied. As we watched, William Price lined up on the sand airstrip and took off. Exactly one third the way along his run, the Beau began to swing. In the desert it didn't matter, and he wasn't in formation anyway. Up, up he soared, tearing away off to the one side. The whole airfield watched, as he fought to level it up. After a minute William Price was in control again, and tootled off about his business.

'I see what you mean,' the flight commander said, thoughtfully.

Around about this time I became very friendly with an Australian pilot called John Bates, known, naturally, as Master Bates. He was shorter than average, with the almost oriental eyes of the Aussie whose pedigree goes back to transportation times. Master Bates was a flight sergeant when I first met him, and the thing that attracted me to him was his way with authority.

The squadron had had six machines on a harbour strafe on one of the islands. We'd lost an aircraft, and it was debriefing time. There was a group captain on the airfield at Mersa Matruh, and he remarked on the fact that Master Bates had brought four of his rockets back. 'You're supposed to fire all your rockets, you know,' he said.

'I was firing in pairs, under orders,' Master said, 'and four rockets gone means two runs at the target. That can't be bad.'

'It doesn't matter,' the group captain said severely, probably piqued at the lack of respect shown to his rank by a Royal Australian Air Force NCO. In the RAF we'd have been terrified to talk back like that.

'Well, if you're so sure about that, maybe you'll come along next time and show us how it's done,' Master Bates said, with his little, perfectly composed smile. He returned the senior man's outraged gaze levelly, and it was the Englishman who turned away.

I was tremendously impressed by this, and set out to get to know the guy. We got drunk together and exchanged confidences.

'Here I am with four operations done, thirty-six to go, or 200 hours, whichever comes first, and I'm with a pilot who can't take off, can't land, wiggles and waggles all over the sky in case the enemy is watching, even though it's broad daylight, and freezes when he's shot at,' I said gloomily. 'There's a whole world full of women waiting to sleep with me when this war's over, and William Price is going to take me away from them, as round as a hoop.' I gazed into my glass of Cyprus brandy, flown in specially from the island where it was sold in bottles that cost more than the drink that was in them.

'Isn't that a very strange coincidence?' Master Bates wasn't a great boozer, but compared to me hardly anybody was. He sipped at his Egyptian beer. 'I have a married Englishman as navigator, and he has me right around the bend and back again, telling me to be careful. Every time we draw near the target he's muttering, "Careful now, John. Take no chances."'

Master Bates sighed. I thought about the sailors who rounded the Horn in sailing ships. When they were up aloft they used to say that one hand was for the ship, and the other was for the wife and kids, and many a barquentine foundered because of it. There's always a time when the only way to see the wife again is to use both hands to make sail, and hold on with your toes. I said as much to Master Bates. He agreed.

'Do you know,' he said, 'Ron has even got a walloping great hunk of armour plating under his seat. He has this fear of getting one right up the jacksie.'

I shuddered. 'I wish you hadn't told me that,' I said. 'That's another fear added to my catalogue.'

7

The empty desert

The old, peacetime airmen who'd done a stretch in the desert called it the Blue. We used the name too, and it was fitting. Somehow, its distant horizon and its barrenness and desolation came together in the word. To me it was always the dried-up seabed, with white sea shells and marine fossils everywhere. The longer we were up the Blue, the more we grew to resemble the Senussi travelling people who traded with us tiny pullets' eggs for tea. Two men could share a tent in complete accord, and barely a word exchanged between them for weeks. Eccentricities grew almost unnoticed. One of the pilots on the squadron took to letting his mouth hang open, waving his fly-whisk and behaving like Edgar Wallace's Sanders of the River. 'My son grows tall and straight,' I heard him reply once to a polite enquiry from Morkas, the camp barber. Morkas was a product of the Cairo streets if there ever was one, and not in the least like Paul Robeson, African chief. We had a CO who went around in a long sheepskin coat, sandals, and only underpants besides. A navigation officer, giving a lecture on the art, began: 'At night-time, go outside your tent and identify Polaris. Get a piece of pointed stick. Put it along the ground with the point facing north. Then go to bed. Next morning go outside, carefully clear the sand away from your stick. Pick up the stick, keep it facing north, walk carefully with it to the flights. Climb into your Beaufighter, hold the stick in the one place. From the direction that the pointed end indicates, work out your course. Then pass the course to your pilot, thus: "The stick points to the cookhouse, so steer over there, where that German tank's lying." That's the correct form of address.'

We had to write all that down, and were marked out of ten on the result.

Ours were not the mackerel-marked, rippling sands of Morocco. There were rocky tufts and knolls, our feet crunched on pebbles. Any given acre of the Western Desert could have been European urban ground, cleared for building. It was the fact that there was such a vast, seemingly endless expanse of it that gave the desert its character. Each windy day I just wanted to keep washing my hair and hands – I came out of the war a compulsive washer – but at night-time, away from the camp, beneath a sky that seemed to be pouring bright stars, that's when I thought of it as the Blue, the distant Mediterranean, the sand and the sky merging into one indigo whole. In the night foxes ran, and gazelles. There were pie dogs and wild cats. If you shone a torch you might hold a kangaroo rat hypnotised. Through the scrub, scorpions rushed away from scorpion beetles. In the year and more that I was in the desert I was never once bored or afraid at night-time. It was the daytime and good flying weather that was bad.

We did a night intruding trip over Rhodes. There was this searchlight at the tip of the island and although I knew it was silly, I fancied that I could see a little fat man walking in front of the beam every so often. The next time we circled near it, I let off a few rounds from my Browning at the beam. I missed, and we were instantly bracketed with enough forty-millimetre flak to bring down a squadron. We tore off back to the North African coast with William Price's words ringing in my ears. 'I think you're lacking in moral fibre,' he said. 'Try to keep your nerve. It's a brave man that keeps out of trouble and doesn't provoke the Germans. After all, we're only trying to starve them out, not shoot them where they sit. Don't be so jumpy on the trigger. From now on you'll seek my permission before opening fire. I'm steadier than you. That's why I'm an officer and you're not. And stop smoking. That's another sign of nerves.'

After five operations we were granted a week's leave in Alexandria, and I couldn't wait to get into the fifteen-hundredweight truck and set off on the long haul down the coastal road. I'd heard so much about the fleshpots from the boys on the squadron, and what was more important, the men I was going down on leave with were cultured. Instead of talking about women, they were looking forward to having some decent, civilised food.

'There's this little gem of a place called "Le Petit Coin de France",' Desmond Garnett said. 'My own favourite dish is roast quail with potato croquettes.'

'By God,' says I, 'but it's one hell of a long time since I put my tooth across the old roast quail and potato croquettes. There was a wee woman in Mackey Street who kept her shop open till after twelve at night, and after the men came in from drinking Tony wine at threepence a bottle, nothing would do them but a feed of roast quail and potato croquettes. Either that or raw tripe, well salted.'

Still and all, I was anxious to break away from the coarse sort of life I'd fallen into since I'd landed in Egypt. Drinking and playing cards and flying. That was about it. I hadn't even bothered to get myself a proper camp bed brought up by the communications flight from one of the shops in Cairo. I was sleeping on a sheet of corrugated iron, set on top of a stretcher, which in turn rested on a pile of bricks. There wasn't much good in kidding myself I had moved up in the world against that sort of background. So I was fully intending to live the civilised life on this leave, and learn all about being a gourmet and a good judge of wine.

I stayed in a *pension* in Alexandria called Essex House. This was one of a number of boarding houses provided under a scheme funded by the millionaire car manufacturer Lord Nuffield, who had a soft spot for RAF fliers. The others in my party were stopping in another place some distance away. I threw my travelling bag down beside my bed and went at once for a shower. After an eight-hour journey down the desert road I was caked in sand. When I came back to my room it was to find a long, thin South African Air Force officer waiting outside.

'Hullo,' I said, rubbing my hair. 'Did you want me?'

'No,' he said, with a sheepish smile, 'I was looking for aniseed.'

'Aniseed?' I shook my head in some puzzlement. 'You won't get any aniseed here.'

He began to speak, thought the better of it, and walked off down the stairs, shaking his head in disappointment. I went on inside and began to climb into my freshly-laundered drill shirt and long trousers, which, under military regulations, replaced shorts at sundown, because that's when the mosquitoes came out. A bite on the bare knee by a mosquito after sundown was deemed to be a self-inflicted wound. So was a dose of the clap, no matter when it was picked up. But that is by the way.

I was standing admiring myself in the mirror, smartly turned out, shaved and shampooed, when a head popped round my open door. 'Sorry,' the head said, and disappeared. I took out my wallet, counted my money, and saw to it that my pay book was tied by a string securely to the left pocket of my shirt, for the Egyptians were fond of stealing pay books. It was rumoured that the German Intelligence paid good money for these, to be used by their reported army of spies all over Cairo and Alexandria. Just as I was locking my bedroom door, a huge, red-haired RAF air gunner came up to me.

'This is room twelve, isn't it, old boy?' He squinted at the number on my door. 'I was given this number by the old lady at the desk.'

'Well I'm afraid I'm in here, and it's only a single room.' I smiled and made to pass.

'What about aniseed, then?' the air gunner asked, scratching his head.

'In the name of Jaze, what's all this obsession with aniseed?' I shook my head and pushed off about my business. The years that I had been in the war had taught me not to linger too long in the presence of headcases. There must be something narcotic about aniseed, I told myself. I wanted no part of it.

On the way to Le Petit Coin de France, I was thinking that the Middle East was certainly a weird place to be. Imagine getting a bang from aniseed. Still, it was better and a hell of a sight more harmless than drinking compass alcohol. We had a guy back on the squadron who simply couldn't give the stuff up. He wasn't very bright, being employed on general duties about the place, and in his innocence he thought that no harm would result from his action if he replaced the special alcohol in the compass base with Egyptian beer. The whole story came out when our communications flight aircraft wandered off track with an empty compass and had to land at an emergency desert strip in the interior. The only people on the strip were three mechanics who'd been forgotten by the Air Force. The three wrote a letter to King George VI, and made the pilot promise to send it over the wireless circuit when he got back, before they'd tell the visitor where the strip was located. On the aircraft's return, the guy who'd stolen the compass alcohol got fifty-six days.

Probably by the time he got out, he was hooked on Brasso.

The lads were waiting at the bar when I arrived. They had a menu, and talked knowledgeably about the French entries thereon.

'I'll take your word for it,' I said.

'We're going in now,' they told me.

'Well, I'll be in in a minute,' I said. 'Don't order for me just yet, I want to wash the sand down with some beer.'

They looked concerned. 'Don't be too long,' I was told. 'This occasion is very special. At last we can behave like civilised British, with good wine, beautiful food and a linen tablecloth. We can forget about the bully beef, the bread full of weevils, the fatty Australian tinned bacon, the water reclaimed from the sea, tasting like soapsuds. It's essential to remind ourselves that we are educated people, and not brute beasts.'

The first Stella beer went down in a blur. The second took a little longer, maybe half a second. Relaxing, I grabbed the third, the fourth. The bar was decorated with a row of little saucers, containing savoury snacks. There were peanuts, little cheese cubes, prawns. I particularly liked the prawns. Some officer from the Navy landed in beside me. We began to talk about Alexandria. He was what we called a base-wallah, dressed impeccably in whites made by a local, very expensive tailor. He wore a gold wristwatch and he went in for after-shave that smelt like aniseed.

'You don't by any chance sniff aniseed?' I asked.

He looked surprised. 'Why not at all, old boy,' he replied. 'Is it any good?'

'Dunno. Just whiffed it off you, that's all.'

'Oh dear me no, old chap. That's my after-shave.'

I told him about the aniseed fiends back in Essex House. We agreed that there were certainly some loose nuts about the place.

'These prawns are a bit of all right,' I said, lifting them by the handful and cramming them into my mouth. 'Before I joined up I'd never heard of frigging prawns,' I confessed. The beer was taking a good hold.

'Never, never say things like that,' the Navy officer said. 'You'll never get your commission if you go around making statements like that. Senior officers may well suspect that one has sprung from the proletariat, but one should never make it easy for them.' My companion squeezed my arm earnestly. 'You have the appearance all right,' he said, leaning back from the bar and studying me. 'You'd make a perfectly presentable officer, dear boy . . .' He broke off. 'I say, you are frightfully keen on prawns, aren't you?' I was whaling into the things, filling my mouth with them as I listened.

'Haven't had a tasty bite for a month,' I explained. 'I'm going into the restaurant to my mates, to get dug into some quail and potato croquettes and good class wine and stuff in a wee minute. Tonight it's Sam the Gentleman.'

The naval officer's arm was around my shoulder. 'That's more like it, Sam. What a nice name, Sam.' His hand tightened on the spot where my diddy would have been if I'd had one. 'We'll make an officer out of you yet. Would you like something with a bit more emphasis to it? How about a cognac, Sam?'

I removed his hand. 'Cognac, sure, Lord Nelson,' I said, 'but don't be thinking my name's Hardy. There'll be no kissing on this quarterdeck.'

I might have been drunk, but I wasn't stupid. 'Well done,' said the sailor, whose name turned out to be Aubrey. 'You carried that off like a real aristocrat.' He ordered the cognac anyway, and we sat for a good hour, enjoying some lively and entertaining conversation.

We were interrupted by Des Garnett. 'Oh Christ, no,' he said. 'We thought you'd gone to dine in some other restaurant. Aren't you bloody starving?'

'I was at first,' I said, 'but I'm all right now. I've had a good stone of prawns washed down with sixteen beers and five cognacs. There'll be time for the roast quail and all that stuff later.'

I fell into bed in Essex House at about three in the morning, after a round of the cabarets with Aubrey. The sound of voices woke me. It must have been about nine. My door opened. A head came round it. 'What the frig do you want?' I growled wearily. 'And don't tell me it's aniseed.'

'There's four of us here, waiting,' the head said. It belonged to a wireless operator of the Royal Canadian Air Force. I sat up. This mystery had to be cleared up once and for all, before I went bonkers.

'What the hell's gates makes the four of you think there's any aniseed here?' I shouted.

Just with that, in walked this number, about seventeen, nearly black, and dimensions to put a man astray in the head. She seemed at home in my room, took off her coat, blouse and skirt, came over to my bed and stroked my cheek. 'Are you first, darling?' Her voice was soft and husky.

There was a wild rushing of feet. Four uniformed fighting men got stuck in the door, struggling to get in. They burst through and landed on my scratcher. The girl was almost buried under their bodies.

'I was first!' the Canadian was yelling. I could see the South African and the other guy from the previous day's visit. Eventually things were sorted out when the girl threatened to go away.

I began to pull my trousers on. 'Whose room is this, anyway?' I was still under the influence of the previous night's carry-on.

'Ze one who sleeps here, he has ze first jump,' the girl explained. 'Zen I let ze ozzer men have jumps. You hurry now. Then ze ozzers can have jump.'

'This is her business address, Room Twelve, Essex House,' the South African explained.

A sudden light dawned. I pointed to the girl, speechless. 'Your name?'

Aniseed stepped out of her knickers, and winked.

Mark Kerr had been sent home very suddenly with his pilot, Len. Seemingly the North Coates Beaufighter wing on the east coast of England was in need of replacements. Our own strength on the Libyan coast had been reduced. One Beau squadron went to Italy, taking Mick McDermott and his pilot with them. They were attacking shipping in the Adriatic and dropping supplies to Yugoslav partisans, the latter a highly dangerous job because of ground fire and high hills. A torpedo bomber squadron of Beaus was to go to the Far East, and thankfully it turned out to be another one. My squadron stayed, moving from one airfield to another, up and down the quiet desert battlefields between Tobruk and Mersa Matruh.

As midsummer approached, our own little theatre came suddenly to life. The Baltimores bombed Heraklion and Rhodes almost nightly as the Allies, on the left flank, took the Italian coastal positions and reduced the number of bases from which the garrisons could be supplied. We had agents in Greece who sent regular reports and from these it was clear that the people on the Greek mainland were suffering fearful starvation under the Germans. If the islands were to be taken it would make a mainland operation a great deal easier. So we attacked the ships and the harbours, and lost aeroplanes doing it. On maybe one sweep in five the incoming formation of Beaus wheeling over base would be an aircraft short, and the watchers on the ground would squint and try to read the aircraft letters, to see who'd gone this time.

It wouldn't have been so bad if the lives had drained away in the sinking of huge battle fleets, or even sizeable single vessels, but our targets were little coasters, albeit armed to the teeth and protected by flak ships and sometimes fighter cover, but still they were only tiddlers in the nautical league. As usual, it was the new crews who formed the great bulk of the casualties. To the keen guy on his second or third operation it was exciting to see a German Navy E-boat, and to put the nose down and try a strafe with the cannon. But E-boats could twist and turn like eels, and they carried multiple cannon themselves. In my time we lost half a dozen crews to this sort of target alone, never mind the flak ships, harbour defences or the fighters.

We had a little Scots fitter sergeant named Dundee who could do a beautiful take-off of Hitler, and this act was helped enormously by the fact that we had a wealth of German Army uniforms to use as props. There was, of course, the ancient ceremony of initiation to the office of Cardinal Puff, involving a ritual of complicated movements carried out in intervals between drinking vast draughts of spirits. There was a sullen, withdrawn group of men who played mahjong in a corner and refused to reveal the rules of the game to an outsider. They would stop playing if even one spectator dropped by. Our squadron doctor was a poker fanatic. I was playing with him one night when he was called to the phone. After a few words, he returned to the game and played for over an hour. Then, pocketing his winnings, he stood. 'Some silly bugger's fired a rifle into his foot, down at the petrol dump,' he said. 'I'll have to go. But keep my place.'

'Was that the subject of your phone call an hour ago?' I asked. He nodded, and went out to treat the poor bastard who'd taken second place to a lucky run at the cards.

Andy McCollum, a warrant officer pilot, was mortally scared of the Italian fighter the Macchi 202. He had almost finished his 200 hours, and the nearer he came to the magical figure the more he talked about Macchi 202s.

'That's a nice drop of Spam,' I might say, eyeing the dinner for the evening.

'Eat it up and enjoy it,' Andy would say, 'for if the Macchi 202s get their way, it's the last bit of Spam you're likely to see.'

If party tricks were in order, I might have been called upon to stand at one end of the mess anteroom, take a run, and clear the bar counter, landing almost face-on smack into the drinks shelves. Another one of my performances was to stand upside down on my head on the bar, the top of my head resting in a hollowed-out loaf of bread, singing 'Danny Boy'. But Willie Wynnersley was top of the bill when guests were being entertained. Willie, a fitter flight sergeant, a man of almost magical skills with an aircraft engine, would ask politely if he might pass the time for the guests with a song. On permission being granted, he would stand facing into a corner, then suddenly turn, holding his charlie in his hand, and sing, 'You are my tart's delight, and where you are she longs to be . . .'

When I left the mess tent I used often to walk well beyond the camp confines, to a steep escarpment. Here I would sit, drunk or sober, in the silver lake that the midsummer moon made of the bare sandy bowl to the south of Gambut. In the distance a pie dog would howl. Occasionally a pair of luminous eyes would gleam close by. The desert had its own fascination for me. I must have been the only man on the squadron who had no sickness for home. I had none of the hatred for the sand and the scrub that the others had. It was a place where regrets had no point. On those nights of peace, I wasn't even capable of concern about William Price and his deteriorating airmanship. That was best left to daylight and the bright blue sky, from where the sun would shine again in the morning, and devour the bare, sandy spaces, and give us perfect flying weather to kill ourselves . . .

In the middle of May a County Antrim Irishman came to take us over. With his arrival, bullshit took first place over everything else. He was tall and thin, a peacetime gentleman flyer who had come in through the Reserve of Air Force Officers. The day he landed I happened to be on standby duty at the flight, ready to take off if a sighting was reported.

'What is the latest from the target area?' he asked me when introductions had been completed. He had most certainly noted my Belfast tongue, as I had marked his own refined overlay on a good Scots-Irish accent, but he'd made no remark. Always alert to negative vibrations, this made me wary. I buried my usual easy-going way with rank and answered circumspectly. A big shipping convoy was expected to try the passage to Crete from Piraeus, Athens. Wellingtons and Baltimores were reconnoitring day and night, and we were standing by against the possibility of a sailing. The wing commander asked me about my operational experience, nodded and went away. I told myself that he was worth watching. He had none of the Irish warmth for a compatriot.

That night the warrant officer who was president of the sergeants' mess was advised that the new CO would like to meet the sergeants. Hurriedly, a party was arranged. All the known characters were asked to be ready to entertain the wingco.

The sergeants had been fond of the outgoing CO. He'd been an easy-going sort of chap, happy to see the work of the squadron done with the minimum of fuss. We'd lived like gypsies, with eccentricities given full reign. The old man had been fond of the odd thrash in the sergeants' mess, and it seemed a good omen that the new man wanted to relax with the senior NCOs on his first night on the squadron.

He arrived with the two flight commanders and a couple of senior commissioned pilots. It was soon noted that he was no fierce drinker. A glass of beer, carefully nursed, did him for an hour. But the visit wasn't twenty minutes old before the 'B' flight commander began to encourage the characters to do their speciality acts. Adolf Hitler was paraded, and soon began to adapt his act to the occasion. When his harangue was at its height and the foam was flying from his lips, he rushed over, grabbed the new CO by the shirt and bent him across the bar. The words 'Irish *Schweinehund*', and 'Orange bastard', could be heard through the torrent of pidgin German. The CO took it well enough, smiling afterwards and nodding his appreciation to Dundee, who had been on the beer for a couple of hours before the arrival of the visitors in order to get himself into a sufficiently Hitlerian mood.

Taff Morgan, the photographic sergeant, suggested a vaulting game. A table was produced, a blanket draped over it. A line was drawn beyond the table. 'What you have to do, sir,' Taff said to the CO, 'is to vault over the table and push yourself far enough beyond to land on the other side of that line.'

What the CO didn't know was that somebody was hidden by the blanket under the table. As soon as the wing commander was heard to take off for his vault, this man lifted the table a good foot higher, and the CO landed on his neck in the corner of the mess. It took about ten minutes to pull him together, but he was in good enough spirits when the star of the evening, Willie Wynnersley, turned to face the audience with his chopper in his hand, singing, 'You are my tart's delight'.

The evening was voted a howling success by all. There was some puzzlement when I declined the invitation to leap over the bar and crash into the drinks shelf, or to stand on my head singing 'Danny Boy', but some instinct told me to be wary of acting the goat in this officer's company.

Next morning the adjutant, to his astonishment, was summoned before the CO at 7.30. This was a full hour before the squadron usually woke up.

'I want the following posted from the squadron as soon as possible,' the new CO said, handing over a list of names. The adjutant glanced at it and whistled. All the characters from the sergeants' mess who had given outrageous performances the previous evening were there. They were on the road within hours. The Irishman had struck his first blow.

Soon we were on parade. 'I am not happy about the discipline on this squadron,' the CO said, 'and I propose to do something about it with the minimum of delay.'

Night guards were reintroduced. There was a parade first thing every morning. Orderly officers and sergeants were required to carry out their duties to the letter, inspecting kitchens and rations, taking guard parades. Training was introduced for the aircrews on standby duty at the flight. Timber Wood, the senior pilot and deputy flight commander on my flight, was asked to give his views on the Browning machine gun. 'It's a thing that's locked up inside the wing of the Beau,' he said, 'and when you push the button the bastard jams.'

But the new man's pride and joy was the barrier pole that he'd had installed at the main entrance to the camp site. He placed two policemen on duty beside it, raising and lowering the thing when a visiting vehicle arrived, even though this only happened about once a week. He had the pole painted red and white. At the entrance to the camp the stones at the edge of the road were also whitened.

As May came to its end, the tempo of operations began to increase. Activity in the Aegean Sea was stepped up. The hitherto slow rate of losses on the squadron began to accelerate. I was introduced one evening to a Canadian replacement pilot called Pat Davies. He was Irish by extraction, and wasted no time in raising the question of the border in Ireland. The beer was in and sense was out, and I came near to flattening Davies at one stage of the evening – not because of the republican stand that he took, but because he took it without knowing what the hell's gates he was talking about. He couldn't name the six counties, even. I left him that night in a boiling temper, stamping white-faced off to my noisy, corrugated-iron bed with my fists balled tight. Next morning I was in formation with the Canadian, when we came across a couple of E-boats escorting a coaster. We tilted over to attack and went down with everything firing. William Price broke off too high as usual, but we got away with it. However, when we reformed Davies wasn't there. He was lost on his first operation. I'd cheerfully have surrendered the border if only I'd left him on better terms.

Another Canadian called Taylor won a DFC for shooting down

a Ju 52 transport. The very next day his was one of six aircraft sent up to attack a small island harbour. It meant approaching from the land. We swept inland and climbed up the slopes of a gentle-looking hill, prior to arriving unexpectedly through the back door. When we sorted ourselves out after strafing the harbour, Taylor was missing. Somebody said that he had hit the top of the hill, an unbelievably silly mistake.

But it was the need to line up for rocket attacks against shipping that was one of the major reasons for our casualties. In order to ensure a hit the rockets and cannon had to be fired in a shallow dive, or in other words right down the ships' gun barrels. At times it was almost too much for William Price. I used to have fantasies sometimes that he was about to bail out of the aeroplane before we ever got near the action.

Formation attacks by night were introduced. Low-level flying around the islands was a tricky carry-on, and several of the navigators had to abort their trips and return to base. On other occasions Beaus just didn't come back and it was likely that they were lost through navigation failure. Here and there the islands had very high hills and it was a highly-skilled business flying below them, trying to see what was in the harbours. Once a flight lieutenant from our squadron found a destroyer during his sweep. The rest of us were spaced out at three-minute intervals behind him. He called up to say that he was going to carry out a rocket attack. Then, still transmitting, he was hit. His screams for the next fifteen seconds or so were the most chilling sound I have ever heard. We could see in the distant night sky his aircraft burning as it plunged into the sea. Not surprisingly perhaps, not a single Beaufighter managed to locate the destroyer when the rest of us came to search the same area. That morning, back at Gambut, we all looked more than usually thoughtful as we sat down to our weevily bread, axle-grease marge and bacon and egg.

On 1 June the big convoy finally sailed from Piraeus. The wing was hastily assembled at Gambut. There was one South African squadron plus two RAF squadrons of Beaufighters, a medium bombing force of Baltimores and a squadron of Mitchell bombers. We had plenty of time. The convoy would take a full day to get to

Crete. A squadron of real live American Mitchells also arrived to take part, but when their cigar-chewing colonel was told that his men would have to sleep on the ground at Gambut after the attack was over, with three blankets to cover each, in accordance with practice, he told his crews to get back into their aircraft, and off they all flew back to Algeria where, presumably, living was easier.

8

Ship-busting

Air Vice Marshal Keith Park, a man who contributed mightily to the RAF's victory in the Battle of Britain and who was banished afterwards, for reasons of jealousy, first to Training Command and then to our unfashionable little war, came up the desert from Alexandria to give us encouragement before the strike. 'There will be severe flak and there certainly be fighter cover,' he said, 'but this is the big one. This will be the Glorious First of June. Go in hard and good luck to you all.'

'Bloody lovely,' says I to anybody who would listen. 'When speeches from military commanders ring out well, start practising the breast-stroke.'

The whole effort was to be led by our squadron. Very interesting. It would, therefore, be led by our wing commander. Even as the CO listened to the recital of odds against his returning, his eye was wandering idly over us, as though he was checking his men for haircuts. I had to give him full marks for coolness.

The scene north of Crete when we hit the convoy and its cover was something to remember. The sky was blackened and torn asunder by heavy, medium and light flak. The ships in the convoy had scattered and their wakes were twisted, tortured streaks on the boiling sea. Ahead of us an Arado 196 floatplane dived for the sea, its starboard float in flames. The float was also the petrol tank. When the attack proper developed, fighters appeared and Beaus jinked and turned frantically. Overhead, in almost stately procession, the light bombers tried their luck on the heeling, turning ships. The first Beaufighter wave was the anti-flak force, strafing the decks of the escorting vessels to keep the gunners' heads down as the second wave attacked the soft ships with rockets. Before the eyes of the whole force, the wing commander's Beau burst into flames. Failing to pull out of its dive, the machine hit the sea, cartwheeled and broke up.

The squadron lost one other crew on the strike and the other Beau squadrons lost two between them. When we got back from the action, we went straight into the bar. Somebody had brought back a length of anti-aircraft steel wire trailing from a wing, and this was laid along the bar as a souvenir. After we got thoroughly tight, we all went ceremoniously out and dismantled the red-and-white barrier pole and peed all over the whitened stones. Then we forgot all about the wing commander.

In the RAF overseas, ground-staff personnel tended to feel a much stronger loyalty to the squadron than did the air crews. This was because the ground staff were required to serve an average of four years before being repatriated. It was a long, long haul, and such men needed something to replace the home link that was available to serving airmen in the United Kingdom who could take leave at frequent intervals. The squadron became their family. The adjutant or the engineering officer or the armament officer became the elder brother, and the commanding officer was the father of the family.

When squadron parties were arranged it was rarely that the idea came from the ordinary fliers, with the exception of those officers who were natural boot-lickers. Parties involving the whole unit were the brain-children of the rankers, and more often than not the festivities were held in the airmen's mess, or the men's canteen in order to emphasise this point.

When I climbed into the rear seat of a Beaufighter to prepare for my first-ever operation, the flight mechanic who tucked me in and fussed around to see to my comfort urged me to put forward my best endeavours in whatever trials lay ahead. 'This is a gen squadron, Paddy', he said. 'You keep up the good work. We've had some bloody good Irishmen, and you look like another one.'

'I'll see what I can do,' I promised.

It reminded me of the time at St Athan when a six-foot-tall corporal PT instructor had strolled into the hut I shared with a couple of dozen other trainee riggers. 'You're Irish, aren't you?' he said, studying me narrowly. I said nothing, following my Uncle Alec's parting advice. But voices all around told the corporal that he'd got it right. 'OK,' said the PTI, 'then you're on the fight bill next Saturday night.'

'I'm not really Irish, Corporal,' I shouted, 'I'm a Scots-Irish coward. I faint at the sight of an opponent.'

But it was no good. That Saturday evening, when I ought to have been getting off my mark with a Welsh nurse who had the biggest headlights in Glamorgan, I found myself squaring up to a guy who was a good stone heavier than me. At the end of the first two-minute round I was blowing like a grampus, my mouth was full of the curious metallic flavour that comes from continuous blows to the nose, and my brain felt as though it had been subjected to twenty-four hours of intensive marijuana-smoking. The one-minute rest had come just in time.

'Hey, Paddy,' my second said, urgently, 'can you hear me?'

I nodded, taking in great gulps of air.

'That guy in the blue corner's winning,' I heard the man say.

'Thanks for letting me know,' I gasped. 'I thought I was battering him to pulp with my jaw.'

'Look,' the second went on, fanning me with the towel, 'the red corner's two fights ahead tonight. Don't go ballsing it up, now.'

'Well, don't tell me about it,' I replied. 'I'm not doing this to myself, you know.'

The call came: 'Seconds out!'

'Remember, Paddy. The red corner.'

'F . . . you, and the red corner too. All I want's out of this with my head facing the right road.'

I felt much the same way where the squadron was concerned, but I wasn't going to tell that to the flight mechanic. The poor bastard had to have something to hold him together. 'Give them Jerries a headache, Paddy,' he said, climbing down the ladder.

When he was on the ground he looked up and gave me the thumbs-up. I did the same. He probably went back to the dispersal tent, full of emotion, not realising that I regarded him as full of bullshit. 'Semper In Excreta,' I muttered, as William Price began to push buttons. 'Bad enough fighting for the Free Peoples of Europe. Now I'm struggling to keep flight mechanics from missing their nooky too much.'

Anyway, the squadron wasn't exactly steeped in tradition. It had begun life as a flight, formed as soon as the war broke out. In 1941 and 1942 it had been up and down the desert in support of the Eighth Army and the Navy, losing Beaufighters and crews steadily, attacking shipping and ground targets in a workmanlike way, earning no special recognition, oiling no particular hinge of fate. No flashy decorations had been won by past members. In the good old RAF way, Distinguished Flying Cross awards to officers had outnumbered Distinguished Flying Medals to sergeants by more than two to one, but nobody had ever won a Distinguished Service Order in the squadron's history. The general morale of the air crews was neither high nor low. The idea was to get the tour of operations over and get the hell out of the squadron to some safe job. There was a good chance with fliers of going back to the UK when the tour was over.

Just before I joined the unit, Churchill, with the same unsound military instinct that had prompted the intervention in Greece in 1941 and the consequent postponement by almost two years of victory in North Africa, had ordered the capture of the German-occupied Dodecanese islands of Cos and Leros. The intention was to impress the nearby, neutral Turks and to persuade them to come in on the side of the Allies. The islands were to be seized by an amphibious attack, backed up by paratroops.

But Churchill chose to ignore the one major lesson of the time – that air superiority over the target beaches was essential to successful landing and consolidation. The Germans had single-engine fighters on Rhodes and Crete, only minutes away from the beaches, whereas the British troops depended for top cover on the Beaufighters, flying from a Cyprus airfield.

In the event the landing failed, the British troops were captured and the squadron lost more crews in its backwater war. By the time William Price and I came to join it, the squadron was made up mainly of fliers who didn't expect to be treated like heroes, and certainly weren't pleasantly surprised as their time spun out.

The matter of gallantry medals did cause some wry shrugs. A sergeant observer on the squadron, before my time, saw his pilot wounded in the legs by ground fire: a considerable amount of blood was being lost, so the observer applied a tourniquet and bandaged the wounds. Then he eased the pilot out of the seat and flew the

Beaufighter home. Over the base he revived the pilot and manoeuvred him back into his seat sufficently for a safe landing to be made. The observer was awarded the DFM. Back home in Bomber Command, with the eyes of the British press on him, the feat would have earned the sergeant the Conspicuous Gallantry Medal, the next thing to the VC.

The officers weren't a lot better off. DFCs were rare, and several were awarded for feats of arms that would have attracted a DSO in the area where the *Daily Mirror* was most active. To rub things in, there was the matter of Group Captain Max Aitken, the famous Battle-of-Britain and night-fighter ace, son of Lord Beaverbrook, the owner of the *Daily Express* and the man who took charge of aircraft production in the crucial days following Dunkirk.

Shortly before I came to the squadron a Fairchild light aircraft landed at Mersa Matruh and out of it stepped the group captain. At the time he was senior air staff officer at our group in Alexandria. Tanned and handsome, he had called in order to brief the men on an operation to be carried out that night. Intelligence had advised that Ju 52 transport aircraft were expected to arrive at Heraklion airfield in Crete, carrying vital stores and troop replacements. After the briefing Max Aitken asked whether he might be given an aircraft and navigator so that he could join in the operation. Permission, naturally, was granted.

Those taking part had been advised to patrol a certain strip along the coast, as it was considered to be the likely route. This they did and they had no luck at all. They came back to the desert and reported accordingly.

Then into the circuit, in the pale dawn light, came the group captain's Beau. He banked steeply and landed. At the debriefing he reported that he had tired of patrolling as suggested, and had boldly joined the circuit over the enemy airfield. It was far and away the most heavily-defended point on the island. Catching the defences on the hop, Max Aitken had destroyed three Ju 52s in the circuit and had damaged two more.

After recommending his observer for a DFC, the celebrated air ace climbed into his aeroplane and flew back to Alexandria, leaving behind a bunch of fliers with egg pouring down their faces, barring, of course, that of the guy who had been promised the medal. There was no doubt about it, in a service made up of squadrons that were just average, ours was a lot more average than most. The only ones really attached to it were the ground crews, and that was only because they *were* attached to it, and couldn't get off it.

Tony Delaney, from somewhere in the heart of working-class Dublin, had an ambition. In the desert he never tired of talking about it. He wanted to be a tramp after the war.

'Did you ever know, Sam,' he used to say, 'that man can live on bread and cheese alone. Therefore there is no need for man to do any work. There's not an Irish housewife in the length and breadth of the thirty-two counties who would deny a tramp at the back door a hunk of bread and cheese.'

He was small and chunky and he had the face of the stage Irishman, with wide grinning smile and all. He held absolutely no rank in the RAF. From the day and hour that he had enlisted in England at the outbreak of war he hadn't been promoted even one tiny step, and the Air Force, more than any other arm of the services, provided lots of little stages in promotion. From aircraftman second class (AC2) it was possible to progress to AC1, then to leading aircraftman (LAC). Having reached that height, it was possible to move to a higher trade group and to start all over again climbing up from AC2 to LAC. And that was before starting in on the stripes at all. But Tony Delaney wanted none of this. He reckoned that promotion knocked the sense of proportion out of a man.

'I can be myself at the bottom of the heap,' he said. He believed that those who truly gave free rein to their natural instincts belonged either at the top of society or at the very bottom. 'In each case, there's no pressure on them to be very different,' he would say. 'It's only the climbers who have to curb their excesses. We tramps and earls are the true nobility.'

He was so outrageous, and so likeable with it, that he got away with everything on the squadron, bar murder. A new adjutant, meeting him for the first time, wanted to know why he was sitting on the fire tender naked from the waist down.

'I'm drying my short trousers,' Tony told him.

'Well, haven't you another pair of short trousers?' the adjutant enquired.

'I left them in a knocking-shop in Cairo,' Tony said, without a blink. The adjutant put him on a charge and Tony got seven days confined to camp.

Tony looked puzzled when the CO delivered the verdict. 'What do you mean, confined to camp?' he asked.

'It means, I suppose, that you can't go out,' the CO said, looking at the adjutant with some distaste.

'Why, where do you think I would want to go round about here?' Tony's gesture took in the whole surrounding, empty desert, the vast, bare stretch to the Mediterranean.

'I don't know where the hell you might want to go, Tony,' the CO said, 'but wherever it is, you can't.'

Tony turned to the adjutant. 'There,' he said, 'see what you've done? Two bloody days on the squadron and you have the CO in a bad temper, and me feeling closed in, in the middle of the biggest frigging desert in the western hemisphere. What other mischief have you up your sleeve?'

He was put to work in the cookhouse as punishment. At teatime the adjutant came round the kitchen. 'Any complaints, men?' he called. Tony raised a hand and stepped forward:

'I have a serious complaint,' he said.

'State your complaint,' the adjutant said, nervously. Already he was reaping a bitter harvest for chastising Tony. The man was loved on the squadron.

'There's a shite beetle in the gravy,' Tony said.

The adjutant closed his eyes involuntarily.

'What sort of beetle?' he asked, for he was fresh out from Blighty.

The sergeant cook cleared his throat. 'Tony means the beetle you get round here that finds little cylindrical pieces of shite and rolls it through the sand to its nest, for grub. Stands on its front legs, pushes with its back legs. Fascinating.'

'Right, right,' the adjutant said, hastily. He turned to Tony. 'And where is the gravy?'

Tony walked over to where the gravy was simmering. 'There,' he said.

The adjutant peered in, his face working in nervous dread. 'Where are the beetles?'

Tony opened the palm of his hand. 'Here they are,' he said, and threw two scaly, brown beetles into the pot. They clung for a moment to the surface, then slowly disappeared.

The adjutant's mouth fell open. 'You, you . . .' He swallowed, went to say something else, then, after a wild glance around at the stoical faces of the cook and his helpers, he fled.

At the kitchen door he turned. 'I hate you, I hate you,' he screamed. 'And as for all the ones who like you – I hate them too.'

The sergeant cook helped Tony to fish out of the gravy the two imitation scarab beetles he'd bought in Cairo long years before: 'I've been thinking over your theory regarding being a tramp, Tony,' he said, 'and for the life of me I can see no fault in it. The old bread and cheese is hard to beat, right enough . . .'

Tony Delaney was my great good companion. I loved to sit in his tent and solve the world's problems. Except for Ireland's trouble. We never once mentioned that. Like me, Tony thought that the desert had far too much dignity to have its air polluted in such a way.

Once, when I was on standby for a strike, I was so engrossed in the conversation with Tony that I actually missed the take-off. The sound of the Beaufighters snarling overhead made me rush out of the tent in panic. The five hours that the aircraft were away seemed like five years to me. I had been down to fly on the strike with a sergeant pilot called Atkinson. I was told when I got to the flights that another spare navigator had been pressed into service at the last minute. A mass of nerves, I counted the Beaus as they joined the circuit. Then I drew a breath of relief. They were all there safely.

'Where the dickens were you when you were required?' the flight commander demanded, when he'd divested himself of his flying gear.

'I got talking to Tony Delaney, and the time just flew,' I said.

It is very likely that no other reason for my dereliction of duty would have been accepted. The offence was a court martial one. My flight commander stared at me for a moment, nodded, and walked away.

Once and only once did I ever see Tony Delaney put out of countenance. The squadron was celebrating something or other and the troops were drunk. In the middle of the festivities a communications aircraft arrived and one passenger was left off. This man, thin, greying, bony-kneed in his shorts, made his lonely way towards the seat of the noise and activity on the squadron. It turned out to be the airmen's mess. All ranks were assembled. Bawdy songs were being sung. At the heart of affairs, as usual, was Tony Delaney. Spotting the stranger, he made his way through the crowd to his side.

'Just arrived?'

The man nodded, diffidently. 'I was hoping to report to the CO. I'm just here on a short visit.' He wore squadron leader's bars on his shoulders.

As usual, Tony ignored the badges of rank. 'You and me are going to sing a song together,' he said, pulling the reluctant visitor to the centre of the huge marquee.

'Give this poor traveller a drink,' he shouted, and within seconds a near-lethal mixture of spirits was handed to the officer. 'Now get that down you,' Tony said, 'and then we'll get going.' With a nervous smile, the guest sipped the drink. Tony put an arm around his shoulder. 'Here we go,' he said, and began to sing:

When you're working down the sewer, shovelling up manure,
That's when the navvy does his bit, shovelling shit.
You can hear the shovels ring, with a ting-a-ling-a-ling,
When you're working down the sewer, shovelling shit, shit, shit.

As the crowd lustily took up the chorus, the squadron leader finished his drink. Smiling approval, Tony at once handed him a second glass. This quickly went the way of the first. Then Delaney swung into the verse: 'Now the foreman said to me . . .'

He stopped. The squadron leader was climbing up onto a table, with arms outstretched. The crowd cheered, then fell silent. The stranger began to sing. His voice, to the appreciative audience, sounded beautiful. He sang in a clear, strong tenor the opening bars of 'The Garden Where the Praties Grow'. The accent was pure John McCormack. The singer was Irish, sounding more Irish by far than

Tony Delaney. Complete silence fell on the audience. Tony stood with mouth open.

The singer tóok the listeners exactly halfway into the song. Then, just when he had his audience lost in a sea of beery sentiment, he stopped dead, looked down at Tony Delaney and smiled: 'You knew I was the padre, didn't you?'

A gasp went through the crowd of airmen. Tony Delaney could only nod, sheepishly. The padre reached down, took the glass from Tony's fingers, drained what was left of it, and then finished the song that Tony had begun:

> Now the foreman said to me, when I came in after eight,
> As he grabbed me by the bollocks and he threw me through the gate,
> If you can't come fucking early then you can't come fucking late
> When you're working down the sewer shovelling shit, shit, shit.

As over a hundred voices roared the chorus, the padre jumped down from the table and disappeared into the crowd. The party saw him no more.

'Blimey, Tony,' somebody said, 'you Irish Catholics are a caution, no mistake.'

Tony reached back and grabbed me. 'That was no Catholic, carrying on like that,' he said. 'He must have been one of these Irish Protestants, like this one here.'

'Well, no matter who he was, or what he was,' says I, 'he didn't half put you in your box, anyway.'

And there was no doubt about it, Tony Delaney was never quite the same outrageous man again. But he was lovely. His time overseas came to an end shortly after that, and one of the first calls he made when he got back to Ireland in the middle of 1944 was to our house in Hillman Street, where he was happy to tell the family that I was fireproof and unsinkable. He and Dad, who was home from sea at the time, spent two days drinking in Jimmy McGrane's pub around the corner. Mother put him up in my bed in the attic, and then he went away, and I've never heard from him since.

Every time I pass an old tramp in town or country, I look into his face closely, in case it's Tony, for he confirmed even to my old dad that that was the life and the only one to follow.

'That wee pal of yours, Tony Delaney,' Dad said, when I finally came back from abroad, 'he talks a lot of sense. Did you ever know that it's possible to live on bread and cheese and nothing else?'

When I had arrived on the squadron it was to find that a cousin of mine was already there. He was Jim, son of my mother's sister and the famous Uncle Alec, he who had left Belfast with me to join the Army at the start of the war. Jim was good-looking, dark, an ex-welterweight boxer who had fought the best back in the Chapel Fields in Belfast, in the days when the second fight on the bill earned a pound for each of the participants.

As kids we had been great pals, but we'd gone our own ways when we were around sixteen. I had been his great fan in his ring days. I was with him once when, after eight sore and severe rounds, he received his pound note from Ma Copley, the redoubtable lady promoter of the Chapel Fields marquee. His manager and trainer took half a crown each as their commission. The two of us set out to walk home, and talk over the fight, which Jim had won on points. We hadn't gone three yards when the manager's brother called from the entrance. Seemingly he expected half a crown too, for being the manager's brother, so he got it. At the top of Alfred Street an old, battered pug tapped Jim for half a crown, for drink. That left ten shillings. We bought a fish supper each in the chippie halfway up the Oldpark Road, and that came to a shilling. We took the fish suppers back to his house and had them with tea and bread and marge.

Jim's face was tender. The insides of his lips were red raw. Both eyes were swollen, and when he took too deep a breath he would wince, for his opponent had been good around the ribs. At about twelve o'clock he said that he'd like to turn in. As I was leaving, he took seven shillings of the nine that were left, and put the money on the mantelpiece for his mother in the morning. Then he went to bed, with two shillings and a fish supper to show for eight two-minute rounds of fierce fighting. Young as I was, I remember thinking that night as I went home that I would cheerfully have stolen for a guy like that, just to see him properly rewarded for his courage and generosity.

He'd joined up in May 1939, before war broke out, so he was a regular, on a six-year engagement. An aircrafthand on general duties, Jim was the corporal on the fire tender. As soon as we could manage it, we went down to Alexandria for a few days' leave. His whole time was spent in restoring order after my many barroom scenes. One of the most easy-going men I'd ever met, he found more excitement erupting around him in a week on the town with me than he'd seen in two dozen ring encounters. But afterwards, in the mornings, he would just look at me, and shake his head, and smile.

Jim's job meant that he would sit at the edge of the airfield when the Beaufighters were taking off and landing. The fire crew hadn't an onerous job on our squadron. Almost all our losses occurred over the target. We had only lost one or two machines around base, because of engine failure.

I arranged for Jim to go up with William Price and me once. We were doing a practice attack on some rocks off the coast. He stood shyly holding onto the hand grips behind the pilot, listening in on the intercom as I gave ranges and firing directions. We were using practice rockets, with concrete heads instead of high explosive or solid steel. One of the Beaus behind us reported that the navigator's intercom was unserviceable, since no ranges were forthcoming, and that the aircraft was returning to base. We drew up at the dispersal beside this machine and, noticing that the navigator's hatch was still closed, I went over, opened it, climbed in and pulled at the flying boot of the navigator, Dibbs, a quiet Englishman.

'Come on down,' I said. There was no reply. I pulled myself up inside the Beau.

Dibbs was lying across the navigation table. There was blood and brains everywhere. A piece of concrete from one of our rockets must have penetrated the cupola and killed him.

If I'd told Jim about it, he'd have had to go in and get Dibbs out. So I had the mechanic phone for an ambulance instead. Jim was too gentle a man to be pushed into such a scene. I was glad that, in all the time he was on the squadron, he wasn't required to risk his neck going into a flaming aeroplane to pull out broken bodies. For such a killer in the ring, he seemed to me every time I was in his company to be one of the least warlike men I had ever met.

Once, when he was drunk, big Jim told me that his stomach was tied in a knot and his nerves nearly drove him daft when I was away on a chancy operation.

'I can't always read the aircraft letters,' he said, 'so would you wave to me as your Beau passes the fire tender? It'll let me know that you've made it OK.'

So I made a little flag, a white one, for peace, and each time that William Price whooshed past the fire tender, too high as usual, I waved my flag out of the open space at the back of the cupola, to let Jim know that I was all right. When my tour of operations was over, my poor cousin was twice the nervous wreck that I was.

If there was one thing that became noticeable in the course of our odd little war, it was the reluctance of our squadron crews to fly unless they were actually ordered to do it. This applied with increasing emphasis as the tour of operations grew longer. There were times when it was necessary to climb into the Beaufighter and get up in the air to carry out duties that weren't going to count towards the magic total of 200 hours or forty trips. On such occasions it was common for the navigator to try to persuade the pilot that there was no point in having anybody in the back seat. Most certainly, after I had completed a dozen operations I took this attitude with William Price. I remember once sitting in the mess playing blind-brag on a sunny afternoon when William Price came on the phone for me. 'I've been asked to scoot down to Edku, beside Alex, and bring back some grub for the officers' mess,' he said. 'It sounds like fun. Get your best dress ready. I'll see you at the flight in an hour.'

'Oh, what a shame,' I said, in a hollow voice.

'Why, what's the matter?'

'I've got Gyppo tummy something desperate, William.'

'But I need a navigator.'

'To get to Ekdu? Just follow the coast. It's money for jam. Whoops! Must go again! Cheerio. Have a jolly good time, looking round the museum, or whatever it is that gets to you. Toodle-oo.'

'Funny, last time I had to go on a buckshee trip you had Gyppo tummy as well.'

'Do you want proof? Want to inspect an arse that's like a sea anemone, then?'

'Have you been to the quack?'

'Have you got a cork bung that'll keep me OK till I get to him?'

'Mm. Well all right. Is there a spare nav about the sergeants' mess, then?'

'Hold on.'

There was always some dim newcomer only too pleased to fly with an experienced pilot, God help us. 'Who's for a wild night in the fleshpots, then?'

'Does this pilot buy the beer?'

'Does he buy it? You won't have to put your hand near your pocket. He's got dozens of contacts with women as well.'

'Right, too true. I could do with a sight of Mossy Face. Cheerio, lads.'

And with a glass of beer in my hand I would sit in my canvas chair outside the sergeants' mess, German panzer hat tipped over my eyes, idly waving my fly-whisk, watching William Price do his sideways take-off in a cloud of yellow dust. With a sigh for lost innocence I would gather up the blind-brag players once again, dealing the cards briskly, sparing one last brief thought for the rookie navigator expecting a night's debauchery in Alex, and in for a nice, healthy walk along the Corniche and an evening at the open-air pictures instead. William Price was a demon for the open-air pictures. He thought that June Allyson was a knock-out. I always thought that the parts she acted were up his street. Tungsten-steel knickers with the combination locked up in a vault in Zurich.

There was a Londoner name of Walker on the squadron, and I could see that he was fairly well under my influence in this matter. His pilot was an Australian, a big man, who wore the bush-hat common to his countrymen. Following my lead, Walker refused to fly on local trips unless he was ordered to do so by the flight commander. There was nothing wrong with his pilot's technique. It was simply a matter of falling in with my theory that there were odds enough against us making it to the end of the tour without throwing another stone weight onto the horse's back for no reason.

Walker's pilot didn't fancy this much. He was a sergeant himself, and right from the start he argued with me in the mess that crews should stay together on all flights. It kept them close, he reckoned.

'Balls,' I retorted. 'Every time we climb into a Beau we're tempting fate. It's like the guy who high-dives into a tank of water and keeps reducing the size of the tank to keep the tension up. Sooner or later he'll end up diving into a wet sponge, and then some bastard'll wring it out behind his back. The tank becomes a sponge from twenty ops on. You do the diving and we navigators'll watch, and give you a bit of a cheer.'

One day the Aussie took his Beau up for a flight test, after an inspection. Walker was sitting alongside me outside the sergeants' mess tent. His pilot was inclined to be a bit on the spectacular side, climbing steeply immediately on take-off, and banking sharply as soon as the wheels locked home. It looked good, and we watched, the pair of us, appreciatively. Then, suddenly, it all went haywire. The bank kept on when it shouldn't have. The Beaufighter turned and turned until it went right onto its back. We were on our feet in horror. The the aircraft went in. It hit and burst into flames only a mile from us. The siren alarm sounded, the shaken Walker began to run, throwing his glass away. I went after him and grabbed his arm and pulled him back. 'You'll do no good out there,' I shouted.

They brought the Aussie back in a sack.

The funny thing was, Walker stopped talking to me after that. He just ignored me completely. They crewed him up with an Englishman and he finally completed his tour and went home, seemingly blaming me in some way for the death of his pilot.

Eric Linfield, a great old cynic, knew that I wasn't bothered one way or the other. 'He's scared of you,' Eric said, wheezing to himself. 'He thinks you're a witch doctor.'

The only time I mentioned the thing directly to Walker happened when he turned his back on me at the bar when I offered a light for his cigarette. I spun him back round to face me:

'Look, Sensitive Sarah,' I said, 'if you really want to purge your guilt over that crash, I'll lay on a series of local flights with William Price, successor to Baron von Richthofen. He's done more to lower morale in the RAF than any flier in the Luftwaffe. Be my frigging guest.'

Walker smiled in spite of himself, and shook his head. But he never did break the ice between us. I think it probably developed into a lucky charm, that silence. We all got funny ideas, in the desert. Still, if that was it, I was happy to oblige.

It turned out that the Beaufighter had a factory fault. There was a weakness in some of the throttle rods. All over the world Beaus had spun in as a result of it. When both throttles were moved forward, the weak one broke, and since take-off was the most likely time for strain to be imposed on the throttles, fatalities resulted in many cases. A good two months before Walker's pilot spun in, one of our squadron pilots had had to make a forced landing near Alexandria, when an engine cut during a mock attack that we were carrying out against the battleship *King George V*, just off the Egyptian coast. Throttle trouble was diagnosed on that occasion, although it was to take months before some analyst back in the UK matched up the crash reports and ordered modification of the part concerned.

Meanwhile, I continued to trim the odds against myself. After a whole series of refusals to fly on local trips with him, William Price one day offered me a bag of white tablets. 'What are they?' I wanted to know.

'Chalk and opium tablets,' he said. 'They'll fix your skitters.'

'You leave my skitters alone,' I said. 'I'll tell you when they're ready to be fixed.'

There was no runway on a desert airstrip, so before a typical strike the Beaus sat in echelon port, with engines idling, awaiting the green Verey light from the tower. Through my headset I could hear William Price's heavy breathing, could hear him go through the cockpit check:

H – Hydraulic power lever on.
T – Trim elevator nose down. Rudder and aileron neutral.
T – Throttle friction – adjust.
M – Mixture normal.
P – Pitch – both levers fully forward.
P – Petrol – tanks full on. Suction balance cock off.
F – Flaps up . . .

. . . Somewhere up in the islands the Messerschmitt boys were doing the same. On board the ships that were our targets the gunners were eyeing the weaving, shadowing aircraft and watching the skies all around. Their multiple pom-poms were at the ready, as the convoy glided through the green Aegean. Along the route, on the islands, the flak regiments watched their radar screens for evidence that the Beaus were on their way. Although we flew towards our targets below the radar range at only fifty feet, so low that the slipstreams left wind-lanes on the sea, the land lookouts would be watching for us from the Cretan hills, or from Rhodes. We could never effect total surprise. Our leading aircraft were picked out well before the attack could be launched.

Stories of the bombing of Germany lead most people to think that the huge flak shells that burst at 10 – 20,000 feet were the worst, but to us heavy shells were like money from America. Even eighty and forty-millimetre shells represented next to nothing in the way of danger. It was the twenty-millimetre and the .500 machine-gun stuff that had far and away the highest killing-rate. Time and again we came back riddled like collanders with bullet holes. Six times at least great chunks were torn out of my Beau by twenty-millimetre pom-pom shells.

Then again, the ships had taken to firing rockets just as we went into the attack. These rockets trailed strong wire cable, moored to deck positions. The wire could slice a wing off like a grocer cutting cheese. As I sat in the cockpit, awaiting the green signal Verey, I would picture the seamen on the target ships, men doing the same job as my father, down below the water-line, waiting for the siren signal that would warn of Beaufighter attack. 'Whispering Death', they called us, because we appeared so suddenly. No sudden could be sudden enough for me, though.

'Boxer aircraft, you are clear for take-off.'

'Roger, tower. Rolling.'

Strapped into the back seat tightly, home seemed a million miles away. There was hardly any thought of it. I used to listen to my pilot's panted fear-signals, and stare at the armour-plating that divided his compartment from mine, and I used to hate him and curse him for his incompetence. I hated his stupid, thin face, the

North-Country note in his speech. I was wound up tighter than the gutty in a golf ball. My loathing for him and his timidity mounted as we waited our turn to take off. Behind the leader as he tore forward across the sand a yellow, fierce dust storm would grow, the number two's slipstream would compound this and, as we shot forward, blind in the sand-hurricane, taking off on a compass bearing only, the tears would come into my eyes as I waited to see whether William Price could get us into the air in fighting nick, or whether he would swing one time too many and send me into the casualty list in *Flight* magazine. The survivors from Cranwell, Mount Hope and Prince Edward Island would see it wherever they were, and say, 'I see old Paddy McAughtry got the hammer. Christ, will you ever forget the time he did the imitation of the squadron leader in the classroom? Didn't know the man himself was at the back of the room, watching him. Some kiddy, old Paddy. Another beer, Jock?'

In the air after a nice, clean take-off, wheeling and banking to join the formation, everything would die down. I would suck in long, deep breaths, begin to do my navigation and radio checks. Always I would crack a joke with William Price before we'd even crossed the Libyan coast on our way to the action. In the air I felt ashamed of the take-off venom. After all, the bitterness saved me from having to pray to God, like all the others. I wasn't worth a frig as a theologian, but it did seem to me that the odds against the Lord giving me a hand were far worse than evens, considering the smoke signals that were going up already from the German Air Force top cover, and the enemy sailors down below on the cold Aegean Sea.

But when William Price did swing, and when, through the driving dust-storm, I could see the aircraft on our port side coming closer, when a collision looked certain, then theology went out of the window and, unheard by anyone except the Lord Jesus, I would scream to him to help me and keep me safe, to spare me, that I might go out and do what, at that moment, I only hoped to Christ was his work.

Then one day, after a perfectly ordinary operation, when the machine hadn't been hit and everything should have been hunky-dory, William Price broke his last Beaufighter. He landed fifteen feet up, we hit the deck like two tons of spuds, and there I was, doing the usual, stepping out of the arse-end of the front half, bruised but cheerful.

They sent him away to a far place. I never saw William Price again. That night I got drunk and ran up one side of the new CO's tent and slid down the other side. I did this many times, all the while thanking him profusely as he lay there with too much sense to reply. Three times I stood on my head and sang 'Danny Boy'. I made forty quid at blind brag. I told the biggest bore on the squadron to wrap up. I broke into the mess kitchen and made a fry-up for six. I was deliriously happy. That night I wouldn't have called the king my uncle.

I was struck down with furunculosis, which is another word for boils and carbuncles. It had something to do with the diet and boozing. My legs were covered in painful eruptions. They were still there when I went on an operation with an English officer pilot, and my legs were covered in bandages. We had a hard time over the target and had to belly-land. When the fire crew arrived to help me out of the aircraft they spotted the bandages.

'Are they bullet wounds, Paddy?' they cried.

'No,' says I, 'they're carbuncles. I felt like shooting holes in my legs to please them.

Soon after, the man who had been navigating for Master Bates left to go back to base and, in a rare moment of sanity, the Air Force put Master and myself together. It was like winning the Irish Sweep. But he was off on leave. I would have to do two more trips with odd pilots before joining up with him.

One of these operations was with a Liverpool man called Ashmore, and one was enough. He was even worse than William Price: holding back when the attack started; weaving instead of going in straight; getting himself cut off from the rest of the attack force, and consequently copping all the available flak to himself. That trip left me a nervous wreck. I was talking about it that night to Peter Green. Peter was Ashmore's regular navigator but the new CO, Dobbin, had borrowed him until a replacement arrived.

'The guy's a bigger menace to the Allied war effort than Goering,' I told Peter.

He nodded. 'It's a relief to fly with the CO,' he said. 'Dobbin's a smashing pilot. But Ashmore? He's a mass of nerves. This is his second tour and he's trying to be too safe. I'm dreading the day when the CO gets his own navigator.'

The day came about two weeks later. One week after that both Peter Green and Ashmore were at the bottom of the Aegean.

I was flying with Clem Fowler on a sweep between Rhodes and Leros, in formation with five others, when somebody called, 'Target, two o'clock. Top cover.'

Over in the distance we could just make out some grey objects on the sea. Above them, circling, were the gnat shapes of naval fighter escort planes. We were flying number two to Butler, one of the flight commanders. He gave us the order to turn. Already the heavy flak was marking the target for us. The Germans, not without reason, were in a high state of nerves.

We were all over the sky. The leader called out instructions as to which sections were to attack which targets. Facing rearwards in my battle station when fighters were about, I hadn't the faintest idea what sort of a force we'd happened onto. Certainly we'd had no indication from either reconnaissance aircraft or intelligence from Greece that any sort of significant convoy was at sea. I was trying like hell to discipline myself. At times like this I used to get the wildest impulse to get out of my seat and run up and down the aeroplane throwing stones, or shouting abuse through cupped hands. The effort required to sit tight and search the sky carefully, taking segments at a time and raking each one comprehensively, almost made me scream.

There was a sudden movement to my right, at the port side of the Beau. Something large and grey swam into view. I couldn't believe it. We had been joined by an Arado 196, the German naval escort floatplane. I'd seen its silhouette a thousand times. This one was alike in every detail. A part of me wanted to look down at my exam paper and note the make and number.

We were joined together for about five seconds in a line of constant bearing. The enemy's starboard side was no more than fifty yards from me. I swivelled my Browning, aimed ahead, and hosed the floatplane with .303 ball and tracer. The tracer showed that it was pouring into the aircraft right where the observer sat, behind the pilot.

Somebody yelled, 'Look out, 109s!'

I should have glanced astern, but instead I kept firing at the little floatplane. I could see both pilot and observer clearly. The pilot gave one urgent glance in our direction and banked to port, away from us.

'Turn port. Turn port!' I screamed to Clem Fowler. He turned and got a burst of cannon in at the target before we had to break off over the ships, an Italian destroyer and one flak ship.

Then we had to go round again and fire our rockets. As far as I could see, the ships were undamaged. The Beaus reformed near the Turkish coast, and the five that were left flew back to Matruh.

I had seen one of the Beaufighters go into the sea. He'd been attacking the destroyer, and just kept diving on when he'd finished and plunged into the sea on the far side of the target. Nobody had been lost to the fighters. Somebody had seen an Arado go down. In view of the fact that only our aircraft had had a clear shot at one, Clem Fowler and I shared the kill between us, officially. I didn't feel terribly pleased about sharing it.

'I suppose it's because those poor little Arados do a suicide job,' the intelligence officer said, when I mentioned my feelings.

'No, it's not that, because I'm not exactly permanent and pensionable myself,' I said. 'It's just that I kept, for some daft reason, aiming at the observer. If I'd tried for the pilot I'd have shot the bastard down on my own.'

I could have done, too. I was close enough. It was funny, how, in the heat of the moment, I'd assumed that the observer was captain of the ship. Maybe it was because, for as much of the time as I could get away with it, I was most certainly captain of mine.

Dobbin, the CO, wasn't too pleased at the squadron's failure to sink the warship. He straightaway organised an outing for us into the heart of the desert, half the aircrew at a time. We sat on three-ton trucks, moving in convoy formation, as though we were ships in the Aegean Sea. Then the other half of the squadron attacked, firing live rockets at a point just beyond the trucks.

One moment there was only the sound of the Bedfords, cruising through the empty desert. Then there was a distant growl, growing swiftly into a roar. Right on the deck, in line abreast, appearing with a suddenness that left the watchers gasping, six Beaufighters lifted themselves up from fifty to eight hundred feet. Then, noses dropping, they fired their rockets in salvo, forty-eight sixty-pound high-explosive missiles, each with the destructive power of a naval six-inch shell. The noise was like a thousand mad banshees. The six aircraft snarled above our heads and were gone. There was nothing, except the clouds of dust and smoke rising from the stretch of sand chosen as the target area. We were left breathless.

'Now you know, between the ships and you, who's boss,' the CO growled later.

It did wonders for our morale. Afterwards, I nearly felt sorry for the Germans.

For all the use that my astro-navigation was to me in Beaufighters, I might just as well have studied the form of the horses. Night-flying over the Aegean Sea was carried out at such a low level and at such speed that the idea of using any navigational aid other than lightning reflexes and eyesight was ludicrous. We went in amongst the islands mostly in moonlight, and the only way we could hope to find an enemy ship was by searching down the moonlane. Radar was useless for this purpose: there was too much land about, islands of all sizes, from towering Crete to bare rocks.

At night the sea was as deep blue as Stephens' ink. If there was a popple on the water we could only just see the white flecks, but they weren't distinct enough to show up on the graticule of the drift recorder. Low night-flying was highly dangerous. The navigation simply had to be right, there was too little room for violent alteration of course at the last minute.

Sometimes I imagined that the sea was greenish-white and luminescent, but this was some trick of the eyes caused, probably, by the strain of intensive searching. The only real relief in sea-searching came when the moon was bright and the moonlane lay like a golden, glittering carpet across the dancing waters. It was impossible not to be impressed by the sight. In a sweep by moonlight there were always spells when the sheer beauty of the Aegean night pushed back the threshold of fear, kept the war at a distance, as the imagination was caught and held by the yellow-gold seaway, stretching away into the night, narrowing with distance, until it met the purple, star-spattered sky.

Once, off Rhodes – and it was the only time for me – I actually saw a ship caught in the centre of the moonlane. It was a small coastal packet. The squadron aircraft were flying at three-minute intervals. At our leisure we circled it, and when the moment was right and the little ship was fair in the centre of the moonlane, we attacked.

The thing to remember at night was to stick to rockets, leave off the cannon-fire. This way, the enemy knew nothing until the banshee wail came down from the black sky.

We scored strikes early on. The ship began to burn, and for a hundred yards on either side of it the cool moon's light changed to the red heat of hell. Then we flew away, and at first light a recce aircraft saw only wreckage floating where the ship had been. And weeks later we learned from the Greek partisans that the ship had been packed with Italian soldiers, loyal to the pro-Allied Italian regime, being taken back to Italy for imprisonment. It was said that two hundred died that night, in the golden light of the moon.

9

Master Bates

Tolerance was low, fuses were set for seconds: steady casualties, bad grub and the desert were combining to make conversation of even the most innocuous kind highly dangerous.

I once stood at the bar of the sergeants' mess when the orderly officer came in, as guest of the orderly sergeant. A glass of beer was drawn for him.

'I don't think much of this,' he said, grimacing. 'It's like soup. Too warm.'

'Well then, bugger off,' says I, 'to where your pampered palate will be better served.'

'You're talking to an officer,' said the guest, who happened to have trained with me in Canada.

'And a bloody bad-mannered one too,' I said. 'You're in here only by invitation, you know.'

'I've changed my mind,' said the other. 'Forget that I'm an officer. Put your hands up.'

I'd stopped putting my hands up before a fight when I was eight. I threw a glass of beer in his face and belted him one across the bracket. Then I reeled as the orderly sergeant tore into me from behind. At this the photographic sergeant thumped the orderly sergeant. In next to no time it was just like a Wild West picture.

In a matter of minutes we all stopped, exhausted. The bar corporal, well used to the charged atmosphere, had pulled the shutters down before the beer had reached the officer's face. He opened the bar again and more drink appeared.

'Give your man a glass of rum,' says I.

'That's better,' the orderly officer said, trying to smack busted lips.

'It hits the spot better,' I said, doing the same. 'How's Irene doing?' I'd met his wife at Cranwell when they were engaged.

'Fine,' said the officer. 'She always asks about you in her letters. I'll be able to tell her now that you're as fit as a flea.'

I was up on a strike against a small tanker close by the island of Kos, flying with a sergeant called Taylor. Upon nearing the target we saw the usual screen of eighty, forty, twenty-millimetre and machine-gun fire. The twenty-millimetre had a quota of self-exploding shells. These were the most unnerving. To see what looked like a wall of high explosive between the aircraft and the target wasn't the best way to beef up morale in air crews. However, on this occasion we flew over-close to a big shell. Something like an eighty-millimetre burst just underneath us. A hole appeared in the floor in front of my feet, another opened in the roof immediately above it, and simultaneously some of the sand that always lay on the floor of the Beaufighter got into my eyes:

'F . . . me!'

'What is it, Paddy?' The pilot was new to the game. His voice had gone squeaky. We were mere seconds and yards away from the point at which the aircraft had to tip over and fly down the barrels of the ships' guns to aim the rockets.

'I'm f . . . ing blind, that's what it is, and I'm only twenty-three.'

'Call the range!' It was my job to watch the radar screen and give ranges to the pilot.

'How the hell's gates can I call the frigging ranges when I'm blind? The eyes are scalded out of my bloody head, you stupid bastard.'

'I'm no good at estimating range . . .'

All this time the German gunners were throwing a day's output from Krupps' armament works at us. The Beau was bucking and leaping and the ether was filled with excited cries from the other crews. Some pilot shouted his call-sign and said he was hit and making for Turkey. 'Gimme the range!' Taylor screamed.

I sat there with my eyes jammed shut and streaming. My ears were tingling from the thunderclap explosion of the shell that had done the damage. 'OK,' I shouted, '1,000 yards, 800 yards – fire!'

With the usual smooth whoosh the eight rockets left their launching rails under the wings and sped downward. I couldn't see a sausage. It was ten minutes before I got my sight back. When we landed my pilot claimed hits on the tanker and, to my utter amazement, the photographs taken from the nose camera proved him right. Nobody afterwards believed me when I said that I'd hit the bastard with my eyes shut.

The journey north to the islands was routine. From a range of fifty miles the coastline of Crete began to show on the radar, and theoretically it was from here that we were within range of enemy fighters, so the navigator swivelled his seat around, faced aft and began to search through the gap in the Beaufighter's perspex cupola. At this point, too, the rear Browning machine-gun was tested. The cannons were cocked, using a lever operated by the foot, and the pilot was advised. The approach to the islands was made at about 200 feet, but for a target of importance the formation would drop to well under 100 feet. As a rule, entrance to the Aegean Sea was made by the western side of Crete and after the sweep the squadron flew southward by way of Crete's eastern straits.

In the bright sunshine the sea was deep blue, turning to green near the islands' golden beaches. We were too busy to appreciate the beauty of the olive trees, descending in dark green ranks to the shimmering bay, or the white jumbles of houses that fringed the little harbours. Our Beaufighters appeared suddenly from the far ocean, and covered the seaside villages with a roar so unexpected that the people in the fields and in the streets were frozen in fear, their heads jerking upward together, their shoulders hunching, their feet fixed immovably to the ground. The maize and the olives and the cherry trees were multicoloured patchwork blurs, seen from 100 feet. There were almond and walnut trees, but at 200 knots we could not see them. The sky was so blue and so bright that every bold line in the desert-brown camouflage of our aircraft was shown up sharply. We were creatures from the African continent, sorely out of place in the islands, an overspill from another war, as much an abomination as the Italian and German soldiers manning the flak guns, and the handful of enemy pilots on the tiny airstrips down below us, on the lovely Aegean islands.

At the moment of passing the edge of Crete, the tension and the watching and the shallow breathing and the narrowing of the eyes began, and it would all last for an hour, until the islands were safely astern. In daylight every cubic kilometre of airspace over the Aegean Sea was enemy sky. When it erupted it did so in a fraction of a second. A bay that seemed harmless and empty suddenly spewed cannon shells, and we could lose a Beaufighter to a flak ship so fast that the crew would have died before the next aircraft in line had time to warn the others.

Then would come the shouting, the banking and turning, the re-forming into attack pattern, the re-entry into the murderous fire. For most, this was almost a relief. No sweep ever involved two attacks. When this one was over it was home James and don't spare the horses. All that remained was to get it over. Go in strafing, fire the rockets, weave as you leave, and piss off to hell's gates out of it.

'Who was it?' the leader would call, south of Crete.

'It was Wyatt,' somebody would call.

'Hey, have a heart,' Wyatt would shout, and the formation would laugh. He was a rare turn, was Wyatt. In a minute we would know that it was one of the new crews.

'Christ, that's rich,' Wyatt would say, and all the way back to Gambut or Matruh the survivors would be rehearsing the stories for the mess that night, and in one day the death of the Beaufighter crew over the sun-soaked island would be staler news than the score in the previous Saturday's Spurs-Arsenal game.

Back on the squadron, after debriefing, after the post-flight meal and the tepid shower under a perforated forty-gallon petrol drum, after three or more Egyptian Stella beers, tasting of onions, and when the tightness had left the chest, then the islands would be remembered, the beautiful Greek islands, the sleepy villages, the broom-covered hills and the cornflower-blue sea. No nicer place in all this world to die.

I spoke over the intercom. 'There are times, Master, old son,' I said, clearly and distinctly, 'when a contemplative smoke in the back seat's hard to frigging beat.' Then I lit up a cigarette and blew the smoke up the fuselage towards him.

'Just as long as it does you good, Paddy,' he said, cheerfully. I then put him to test number two. I switched on the Armed Forces' Radio. The music filled his earphones as well as mine. It used to give William Price the screaming willies. 'How can you concentrate on your job with that rubbish in your head?' he used to complain.

'D'ye know this?' Master Bates said, after a minute's listening. 'It's time Vera Lynn jacked it in. She brings my morale to rock bottom when I hear her whining voice.'

The final test was to come. Just outside range of the holy murder that was being kicked up by two German ships, as we were preparing to go in for our attack, I spoke.

'What I want you to do here, Master,' I said, 'is to go straight in, knock the shit out of them, and come out doing loops, rolls, inverted flying, and Immelman turns. OK?'

'Your wish is my command,' he said. I gave him the radar ranges on the way into what looked like the bowels of Vesuvius in full eruption. At 800 yards I yelled 'Fire !' The eight rockets went howling towards their target, and I cursed and blinded at the top of my voice to get rid of the tension as we twisted away to the safety of the Turkish coast.

On the way back I brought him up a drop of tea out of the Thermos, and stood behind him smoking and criticising his flying. 'You'll have to work on that throttle,' I said. 'We were only doing about 300 knots going into that target. I want it up to 3,000 as soon as possible.'

On the last half-hour of the run home I sang, 'There's an old Irish lady I'm wishful to see,' and began to tell Master Bates about my people back home, and about how an uncle of mine owned a shipping company once, and drank every halfpenny of it. I always reserved lying of a high quality for the people I admired most.

On that trip, for the only time in my operational experience, I almost made a bad landfall. We had to make a sharp alteration a few miles off the airfield. It bothered me no end, and afterwards I carried out a thorough inquest into my navigation. I didn't want Master to have any doubts about me.

'I found out what caused that poor landfall,' I said over a beer the next evening. 'It was deviation in the compass, caused by that armour-plating ball-protector that your previous navigator put in.'

'Are you keeping it for your own protection, then?' Master asked.

'No,' I said, 'I threw the armour plating away. It doesn't really matter about my clinkers. Nobody seems to be interested in them anyway.'

We flew merrily together, Master Bates and I, and the hours rattled up satisfactorily. Around us at Matruh and Gambut, for we kept shuttling back and forth between the two in the craziest way, our little force was depleted even further. The Baltimores went away, as did the Wellingtons, and we were left with only two squadrons of Beaufighters. Up in the Dodecanese, the single-engine fighters seemed to have gone, much to our relief. Our job was to contain the garrisons, to show force, and to sink their ships, God help them. The big convoy of 1 June had been the last, brave effort. All that the German troops had to keep them supplied were the transport aircraft and a motley collection of little coasters. Thanks to the underground resistance in Greece and on the islands, we even knew the names of these ships, and had their silhouettes up on the operations room wall.

At about this time we had a visit from a celebrated stage magician of the time, Jasper Maskelyne, from the great team of Maskelyne and Devant. The authorities had had the idea of using his illusionist talents to conjure up whole fleets of cardboard and canvas aircraft, in an attempt to fool the Hun as to where the invasion of Europe was to take place. Whether the huge numbers of stage aeroplanes that were scattered about Gambut and Matruh for the edification of the recce Ju 86B that came over every day fooled the enemy I never did find out, but Jasper Maskelyne seemd to grow fond of the desert, for even after the invasion he turned up to visit us. One night he demonstrated for my particular benefit the art of swallowing a razor blade. Into his mouth it went, the great man chewed it up, took a drink of beer to wash it down, and opened his mouth to let me see that the blade had gone.

Well, so it had, but his tongue was cut to bloody pieces. He might have been good at the camouflage, but he was the worst razor-blade swallower in the business. 'It's a good thing,' says I, 'that it wasn't a sword.'

Since the time was coming when the garrisons on the islands would be thinking of surrender, we began to help them along this road by attacking barracks and gun positions with rocket and cannon. Thus, despite the dwindling importance and scope of our little pocket of the war, the authorities back in Alexandria continued to dream up ways by which our casualties could be kept to a steady and worrying rate. Since I had joined the squadron, almost a squadron's strength had been lost. Whether this shortage of fliers was why I was asked to appear before an officers' commissioning board I don't know, but summoned I was, and they couldn't have picked a worse date. The night before the interview the rains emptied down and my tent, which had never had a protective trench built around it because I'd always been suffering from hangovers, was flooded right up to my corrugated iron bed. I had to borrow somebody else's uniform, and it hung on me like something out of a jumble sale.

The board was held in Tobruk. In a room, waiting, were a couple of English pilots. They were talking about their fathers' occupations when I arrived. 'My old man's a butcher in Stockport,' one said, 'and I work for him. I'm going to tell the board that I'm a meat provisioner.'

The other guy was already practising an officer's accent. 'I was only a bloody clerk,' he said, 'but I did a bit of figure work. I'll tell them I was studying to be a chartered accountant.'

I was first in. The door stuck, then burst open suddenly when I heaved on it with my shoulder. Mouthing oaths, I arrived all of a rush in front of the board. One member was my flight commander, another I didn't know, and sitting in the chair in the middle was the great war ace Max Aitken. They were smiling.

'Where on earth did you get that uniform?' my flight commander asked.

I grimaced. 'I was flooded out,' I said. 'This is big Cresswell's outfit.'

'Don't you think that, for an occasion like this, you should have searched around for something your own size?' the third man asked.

'Big Cresswell's the only one who keeps his uniform pressed, big and all as it is,' I explained.

They got down to talking about my flying career. At that stage I'd done about 150 operational hours. 'How do you feel, stuck up the desert in a war that nobody seems to be interested in?' the same man asked.

I shrugged. 'I've done worse that this in my time. It's better than peeling spuds.'

We chatted about flying. After a while the chairman indicated that the interview was over. Thinking of the two candidates outside, I showed disappointment. 'Aren't you going to ask what my father does?'

Max Aitken smiled. 'Please tell us.'

'He's a donkeyman on a merchant ship,' I said. All three were fascinated. They'd never heard of such a thing. Enthusiastically I explained about the hierarchy in an engine room: the progression from trimmer, wheeling the coal to supply the firemen; to the firemen themselves, feeding the furnaces; and, beyond the stokehold, to the three grades of greaser, of which donkeyman is the highest. 'He's a leading hand,' I said, 'and the name comes from his perk in port, when he alone is allowed to light one fire and run the donkey engine off it for loading or unloading the cargo, with all the overtime that's attached to the work.'

'Do you know,' Max Aitken said, 'I don't think I'd mind being a donkeyman for a while.'

'Well, maybe not,' I said, 'but my dad has been torpedoed twice. I think I should give you the bad news.'

Everybody was laughing when the interview ended. When I joined the other two they wanted to know if I'd been asked about my father's occupation.'

'I was,' I said, 'but it held no terrors for me, that one. My old pater happens to own a shipping line.'

Three days later Master Bates was awarded the Distinguished Flying Cross for resolution in carrying out attacks against heavily-defended targets. He deserved the DSO, but the Beaufighter boys always did get the short change.

The same day that Master got his medal I was made a pilot officer. They threw a party for me in the sergeants' mess and, just to show them that being an officer hadn't changed me, I fell out with some pilot from Newcastle-on-Tyne and the two of us left the mess furniture in matchwood.

What really put me in a bad temper that night was the fact that I was told by another officer that they had commissioned me rather than give me the Distinguished Flying Medal. I'd far rather have had the medal. Later, in Greece, I came upon a general instruction to commanding officers: 'Many cases are being reported of NCO aircrew being recommended for commissioning as an alternative to the award of the DFM, and this is producing some officers of indifferent quality. This practice is to cease forthwith.' The Air Ministry probably had me in mind when they issued it.

One day four of our Beaus were in the central Aegean carrying out a sweep when they were jumped by Seafires from a nearby British carrier. Luckily all our pilots were experienced, otherwise there might have been severe damage done. The Beaus made for Turkey, this being a favourite dodge on our part when the Huns made an interception. The Turks practised benevolent neutrality towards the Allies, meaning that they left our aircraft in peace to fly anywhere they wanted around the coast whilst dodging German fighters, whereas the Germans got the full works from the anti-aircraft batteries. When our machines returned that day a complaint was made to the Naval HQ at Alexandria. With the complaint went a sniffy comment on the general ability of the fighters concerned.

A few days later I was in Alexandria on leave with Master Bates. I arrived fully resolved to have that outstanding date with the roast quail and potato croquettes. So far I'd made two attempts at it, but always at the bar some guy turned up who kept me talking and filling myself up with prawns and roast peanuts. 'You're an officer now,' Master Bates said severely. 'You'll have to learn all the fancy ways.'

In his case, being an Australian, he didn't need to learn fancy ways. The British assumed that Australians were unteachable in such things. Master himself, who, unlike the many chancers at the time claiming to be chartered accountants, really was one back home in Sydney, could turn on the sophistication or switch it off, as the occasion demanded. On this leave he was setting out to educate me. 'You'll get into bad trouble, walking past the military police over and over again, to see them salute you,' he cautioned, following an episode in the main street.

'Oh, is that so?' I retorted. 'Well just see this.' I returned to the two MPs I'd been tormenting and stood before them. 'You two bother me,' I said. 'You have an insolent look about you. Don't think I haven't noticed it. I'll be back this way soon, and if you haven't lost that look, by heavens watch out.'

The ramrod-straight soldiers stared dead in front and made no reply. On my turning to go they saluted with parade-ground snappiness. I returned their salutes and took a couple of paces away. Then I turned suddenly. One was mouthing to the other some word that began with 'f'. I raised a finger in warning, waited for a further salute, and was dragged off to Le Petit Coin de France for schooling.

There were Master Bates and myself and an alcoholic South African pilot, who had to put his hanky round his neck and hold the end to keep his glass of whiskey steady at breakfast. Up to the bar we went, and were handed the menu. 'There's no need to give me that,' I said, 'for I know what I'm having. Wheel out the roast quail and the spud croquettes. Tell me when they're ready, and leave the rest to me.'

We hadn't been there five minutes when in walked this party of naval officers, nice and neat in their whites. They stood near us at the bar, talking in the fruity voices that naval officers adopted then.

'Well shut my mouth,' says I, 'if they're not Fleet Air Arm.'

Indeed they were. They all wore pilot's badges. I nudged one of the newcomers: 'You weren't by any chance up around Kos and Leros last week, were you?'

Their mouths all fell open at the same time. They nodded.

I tapped my chest, pointed to Master. 'Beaufighters,' I said.

'Oh, good God.' Aghast was a suitable word for the occasion.

I slid off my stool. 'Do you think we look like frigging Germans?'

'I say, old chap, you're not going to be nasty, are you?' One of the pilots, a thin beanpole, popped a peanut into his mouth and stared at me with enormous interest. His companions were equally composed, against the threat of busted noses. It suddenly dawned on me that I was into a new scenario here. Up until then, for all of my life, when I'd wanted a fight to break out, the other half of the cast had generally been happy to oblige. Here we had a situation where words were expected to do the fighting. I considered my position.

'Oh, I say! It's Sam, isn't it? Dear old Sam. However have you been?'

It was Aubrey, looking absolutely gorgeous in his tropical whites. He took my elbow, stared at me with his head back. 'Never! They've never made you an officer! You shouldn't have let them, dear boy. They've utterly ruined you.'

'This is Aubrey,' I said to Master and the South African. It was all one to the South African. He was already heavy in the eyelids.

'Sam is lovely and hot-tempered,' Aubrey explained, putting his arm around the South African.

'What do you mean, ruined me?' I sank a quick brandy. 'Wasn't it you that was giving me all the tips on how to become an officer the last time I saw you?'

'Yes, but you shouldn't have paid any attention. You were a delightfully rowdy sergeant.'

Somebody tapped me on the shoulder. It was the oldest of the Navy fliers, a schoolmasterly man with wise eyes: 'May I buy your party drinks?'

'You've talked me into it.' We took over a large table, Aubrey acting as chairman. The waiter came over. 'Nine beers and nine brandies,' I shouted, 'and we want no arsing about. Keep hitting this table with drink every fifteen minutes or we'll wreck the bloody kip.'

Sometime or other during the course of the evening the waiter came to tell Master Bates, the South African and me that our roast quail and potato croquettes were ready to serve. We paid for it and stayed where we were. The whole lot of us stuffed ourselves with prawns and roast peanuts, while commiserating with the ones who were undergoing a court of enquiry for trying to shoot Beaufighters down. At midnight we decided to go to a cabaret. 'Are you coming, Aubrey?' I asked.

'Shan't bother, Sam,' he said, helping the South African to his feet. 'I'm taking this poor chap back to my place.' As they left, with arms round each other, Aubrey turned to me and winked. He was a great character. I could just see him on the bridge of a frigate, immaculate, unhurried, in the face of a cruiser attack. 'We're going to play Beaufighters and Seafires,' he said.

I never saw Aubrey again. But I knew which of the two would be shot down.

The Navy sent a fighter direction ship, with naval escort, north of Crete to attack the nightly air-lifts coming from Italy to the islands. Our squadron had been engaged from time to time in trying to intercept the Ju 88 and Ju 52 traffic over the enemy airfields, but results were negligible. We had only the rudimentary ASV radar equipment to help us, and this was really only a navigational aid. It would give a blip for a coastline, but as an air-to-air weapon it was almost useless. We had to fly right over the flarepaths to get a night sighting, and this was an extremely dangerous manoeuvre, one that had cost us a number of crews.

The fighter direction vessel turned out to be the one-time Belfast-Liverpool ferry *Ulster Monarch* and Master Bates and I were among the first on the squadron to give top cover to it and its escort during daylight hours. I was so excited flying over the ship, on which I had sailed back off leave several times, that when the head of operations on board asked our leader if he could extend his stay until dark, and the flight commander refused, I lost the head. 'What the hell's he talking about?' I shouted to Master. 'If he doesn't want to stay till last light, we do.'

'Oh, do we?'

'Too Irish we do. We're all right for fuel for thirty more minutes. Tell the leader.'

As a result the *Ulster Monarch* was protected until it was dusk. Nothing happened but, as I said to Master, you never knew about these things.

In the mess that night Master told the others what had happened: 'If it had been the Dublin-Liverpool ferry he wouldn't have given a monkey's,' the Aussie said. 'But because it was the Belfast boat I had the impression he'd have dived overboard if necessary and swum round it with a knife in his teeth.'

In three evenings the *Ulster Monarch* vectored night-fighter Beaus, fitted with the latest air interception equipment, onto a stream of troop and cargo-carrying Ju 52s. The night-fighters didn't even have their paint scratched. Later we had the teeth-gnashing experience of seeing long lists of decorations announced for those who had been directed by the ship clean up the arses of the Germans. Pupil pilots flying Tiger Moths could have shot the poor bastards down. For protecting the naval convoy for hours at a time, in daylight, two hours' flying time from base, all that we got were bloodshot eyes.

With the advance of the Allied armies northward in Italy, the isolated Greek and Aegean garrisons were given lower priority. The force operating from the desert against the supply lines to the islands was reduced to three squadrons. British naval units began to venture into the area for the first time since the German invasion of Crete in 1941. We were given the job of escorting them. In December 1944 Master Bates and I carried out a photo-reconnaissance of Rhodes harbour in daylight, prior to a bombardment by the cruiser *Black Prince*. It was hair-raising, the more so because we were almost finished with operational flying. Master flew lower on the approach to the strongly-defended harbour than I'd ever thought possible. He operated the nose camera and I took a couple of shots with a hand-held model out of the back window. We rocked and weaved out of the harbour area, took about two stone of lead on board, and made it to the quiet country beyond the town. We suddenly came on a large party of soldiers bathing on a beach. They scattered, but there was nowhere for them to go. We raked them with cannon fire and as we left I opened up with my single rear Browning.

Of all the memories that I brought back with me from the war, this one goes deeper than any other. I kept thinking then, and still do, that there was no need for it. Those soldiers didn't pose a military threat. Somebody has suggested since that they posed a considerable threat to the Greeks on Rhodes. Maybe so, but I'd rather not have seen the men so clearly as they died. I'd never have made a foot soldier.

They took me off operations when I'd done thirty-seven trips and 180 hours. In Air Force terms it was average. But I was happy with myself. I'd beaten heavy enough odds. Half of the group who had posed with me outside the hangar at Catfoss were dead.

I went down the desert road for the last time with a warrant officer navigator called Clegg. He had trained with me in Canada. A quiet chap of my own age, he had come up to the officers' lines on our last night to escort me to the sergeants' mess, where they were anxious to give me a goodbye party. Already high from the hospitality of the officers, I asked Clegg to wait in my tent while I replenished my cigarette stock. By then we had begun to get the British brands as our ration, instead of a South African variety of lung-burner called 'C to C', for 'Cape to Cairo'.

He sat on the other bed, watching me. Full of euphoria, I pointed to a Primus stove just by his feet. 'There's a story connected with that stove,' I said and, giggling, I told him how the Australian Dunne, long dead, had placed a tinful of Friar's Balsam over the Primus, waited till it boiled and gave off fumes, then squatted over it to let the steam get at his piles. He was holding his sexual equipment in the one hand and using the other hand to balance himself on the bed, when I offered him a smoke.

'Without thinking,' says I, 'he let go of his clinkers to take the fag, they plopped into the boiling water, and Dunne's head frigging nearly went through that tent roof there. I've often heard the word anguish used, but that's the only time I've ever actually seen anguish on a man's face. Still,' I finished, 'he never got a chance to see if his tentacles were affected or not.'

'No, he went in by Leros,' Clegg said. 'I was there, and it's a thing I don't think it's right to joke about.'

I straightened up. This was the first time I'd ever seen any sign of strain about Clegg. If ever there was a stoical, lower middle-class Englishman, he was the one.

'You sound just like a man who's been flown to exhaustion,' I said.

'Well, one thing that's good about it is to get away from ignorant bastards like you,' he said. 'What sort of life did you have in Civvy Street, for Christ's sake? You act as if you've just been up for a bloody joyride, instead of getting the tar cut out of you for thirty-seven trips. What the hell sort of person are you? Have you no sensitivity at all?'

I could only stare, at first. Suddenly he seemed to have stripped away from me all the protective clothing I'd grown since I'd walked off the Stranraer boat four and a half years before. I was an ignorant gulpin from the back streets of Belfast, and had never been anything else. English grammar-school and public-school boys felt sensitivity in a war, but rough, coarse, Irish slum products just took it all in their stride. I really should have joined the Royal Ulster Rifles and gone to earth in the trenches where I belonged. What the hell was I doing in company with the Cleggs of this world? I had a native servant, for Christ's sake, when by rights I was only servant material myself.

I simply couldn't answer. Tipping my hat onto the back of my head, I walked out of the tent and led the way to the sergeants' mess. Clegg had ruined my last night. I couldn't get drunk, no matter how much I tried. The lads in the mess were disappointed. No 'Danny Boy'. No rows or fisticuffs. I found myself talking in a corner of the mess to Chris, an American who was unfailingly polite and listened beautifully.

'Clegg just told me tonight that I'm an ignorant bastard because the ops didn't get me down. I'm as low as I've ever been. Is he right, for Jaze sake?'

'Look, Paddy, just think about it. You've never given a damn what people thought about you, have you? You've done what you wanted on the ground, and you've been nobody's man but your own. You ignored the rules if they didn't make sense to you, and instead of being punished you've been awarded a goddam commission!'

Chris was a warrant officer pilot. He had a brother in the RAF who was a flight lieutenant and was in the Middle East too. It made things just a little awkward when they met, since officers tended to use certain bars and restaurants, more or less exclusively, and NCO guests were the subject of cold and inhospitable glances. But Chris couldn't get his comission. It was a fact that surprised many, for his surname was Westinghouse and the family name was known all over the world, wherever electrical appliances were sold.

'The Irish have a special dispensation in this world, because they're good scrappers,' Chris said. 'But in general the people who dislike this arrangement are not big buddies in a jam. Clegg doesn't strike me as any kind of a wildcat.'

It made me feel better, but I still went full of doubts to my sheet of corrugated iron on my last night in the desert. I even forgot to take one last look at the night sky. It was the meteorite season too. Unforgiveable.

On my last night in the desert I didn't go out by myself, as I'd done so many times, unknown to any of them, and sit at the edge of the escarpment, in the moonlight. It was here, where the sensitive ones like Clegg had never seen me, that I'd taken my terror out, recognised it, pissed on it and put it back.

Next morning I was offered a lift in an ambulance. The journey to Alexandria was 400 miles. An ambulance offered undreamt-of comfort. Clegg made to climb inside, to where I was stretched out in a bunk. I pushed him back.

'Get up with the driver,' I said. 'You might be a more precious cargo than me, but I've got the rank, so up yours.'

10

The war is over

Master Bates went back to Sydney. We had a monumental thrash before he left. On the way back from a cabaret, where six of us were ejected for blowing up French letters and letting them swoosh down from the balcony onto the stage, two stupid Egyptians tried to rob us. They were totally inexperienced at the game, giving their intention away far too soon.

'Would you like nice schoolteacher?'

All the whores were nice schoolteachers, of course. I had even been told by a hard man from the Eighth Army about an old Egyptian bint who plied her trade while lying with her head in a dog kennel. 'She claimed to be a lecturer in English at Cairo University,' the soldier said. 'Her pimp told the customers that she was too well known to reveal her face. In fact, she looked like a bloody hippopotamus, in the face.'

'And how do you know,' I'd asked, sharp enough.

'Because I pulled her out and looked at her, after,' the soldier said.

'Go away,' said Master Bates and I to the slim, twenty-year-old man offering us the schoolteacher. Drunk as I was, I saw his eyes dart to a spot behind us. Another man, bigger, was standing in the darkness. He lunged forward. As Master, who was much lighter and more delicately built than I, moved to take the pimp, I turned and let the other one have a boot in the clinkers. When he doubled up he got the same boot across the side of the head. We took off, the pair of us, roaring with laughter, so close together after the desert that we'd have died for each other. Come to that, many a time we bloody nearly did. Later, in Master's hotel room, we went back over our trips together. The most risky operation of all was one in which the idea was to blow up a cargo ship called the *Drache* lying in Port Vathy, on the island of Samos. What with the wheeling and turning that went on on the way into the place, we ended up going in last of all. We attracted the undivided attention of every gun on the island, and came back riddled with holes, and with half of our tailplane gone. 'Some day,' said Master, 'I'd like to go back to Samos. It looked such a lovely place.'

I was sorry, too, that I'd not be seeing the Greek islands again. And of all the islands strung out in the Aegean Sea the one that I too liked best was Samos. I made a promise to Master Bates that one day I would go back and drink Samos Sec with an old-timer, by the quayside.

Dedalus and Icarus, his son, flew over Samos wearing wings of feathers and wax, but the wax on Icarus's wings melted and he fell into the sea. A good many other aviators fell into the same sea in my time. Pythagoras was born there. He was the first one to realise that the explanation of the universe was to be found in numbers. As a navigator, I had relied for my very survival on an ability to get on amicably with numbers.

'Zeus and Hera went to Samos for their honeymoon,' a classical scholar told me once, as we both sat at the flights on standby. 'It lasted three hundred years.'

'Three hundred years, fancy that,' I murmured. 'I make that 15,675 copies of the *News of the World* lying beside the bed.'

After all the goodbye parties there wasn't much left in the line of stage farewells. I just shook hands with Master Bates, slapped him on the back and walked away to my own hotel room. When I got there, and sat down with a beer in my hand, I thought about the people who hadn't made it to Alexandria. People like Skippen and Truscott.

Truscott had been all the way with me through London, Cranwell, Canada, Catfoss and the squadron. He was quiet, well-mannered, like Skippen, his pilot. Somehow, as their hours piled on, they never acquired the roughed-up appearance that the other survivors wore. For some odd reason, too, their flying lives seemed to be charmed. It soon became a squadron saying that if Skippen and Truscott were there in the formation there'd be no kites lost. Then came the sighting of the SS *Orion* in an inlet on the coast of Dhenousa island.

There were eight of us. We took a look, flying out of range of flak.

The little ship was jammed tight under a steep cliff. There was only one approach, and we couldn't get at it in line abreast, the usual form of attack. It would have to be line astern.

Truscott had always had this regret that he was never treated as I was around the squadron – like an old hand. 'You've seen so much action,' he said. 'You look like a real old hand. I wish Skippen and I could be old hands. We could wear German hats, like you.'

We were all cut to ribbons going in to blow the *Orion* up. The ship's gunners couldn't miss. But the Beaufighter was one of the toughest aeroplanes ever built by any nation. So long as the pilot wasn't hit, or an engine, there was a chance of coming out. We only lost one on the *Orion* job. Skippen and Truscott. After over 100 hours' operations, too.

They'd always wanted to be old hands. Well, they went into the side of a cliff on Dhenousa. They were old hands, all right. As old as the hills.

The two or three days that I spent in Alexandria, waiting to go to Greece, were full of morbid musing of the same sort. I was bloody glad when I jumped into the back seat of a Beau in February 1945 to navigate from Geneacilis to Hassani airport, near Athens. It was my last trip in the Beaufighter. From then on I was to be a penguin as we called it, doing a ground job.

I had flown 330 hours in Beaus, 260 of them with one of the worst takers-off and landers in the Air Force. It wasn't perhaps surprising that, when I arrived in Greece, I jumped ship, took off without leave and went native, went on a bender that lasted two weeks. I lived with the Greek dockers in Piraeus for some days, sleeping in bars, then I took up with some airmen from an adjoining squadron, lived in their billet, swam and drank with them. From time to time word would be brought to me that the RAF would like me to return to my unit, but I let the bender run its course. Eventually I reported back for duty, received a shellacking from the air commodore and was given a ground job, as transport officer on a big base located in a suburb of Athens. I had no sooner settled in there than I went down once again with multiple boils and carbuncles all over my legs, good old furunculosis. Apparently the condition was brought on by poor grub and self-abuse. The story of my life. Put music to the words and you had my signature tune. I was removed for treatment to hospital.

General, later Field Marshal, Alexander inspected the military hospital during my stay there. The officer patients stood by their beds, waiting to meet the great man. I was the only RAF man in the ward. All around were soldiers who had been wounded in honourable combat, infantry subalterns, sappers, army gunners. As each one chatted to the general my heart sank lower and lower. This one had been hit by Mauser fire during the storming of a farmhouse, that one winged by grenade shrapnel. Lovely, bloody lovely. McAughtry's war was about the bring the house down in gales of laughter. The great man appeared in front of me. A crocodile of about twenty bigwigs trailed behind him. 'Ah,' said General Alexander, 'the legs, I see.' I said yes sir, yes it was the legs all right.

'And where did you get it?'

'Mersa Matruh,' I said. The general was greatly surprised. It had been over two years since there'd been fighting at Mersa Matruh.

'Still troubling you?' he asked.

'Well, it only broke out this week,' I explained.

Alexander looked puzzled. 'What only broke out this week?'

I pointed. 'The legs.'

'You were hurt in 1942 and it's only breaking out now?'

'No, I wasn't actually hurt, sir.'

'What precisely is the matter with your legs?'

He didn't exactly add 'for Christ's sake', but the words hung in the air between us.

'Furunculosis.'

General Alexander looked helplessly at a red-tabbed Medical Corps colonel nearby. There was a whispered conversation between them.

'I say, bad luck, old chap,' the GOC said to me.

By this time I was so embarrassed and confused that my mouth ran off on its own. 'They're all covered in carbuncles. Something frigging desperate,' I said, in a dreamlike state.

He suddenly smiled. 'Where are you from?'

'Belfast, sir.'

'I thought so. Well, it wasn't very nice, roughing up shipping and harbours, but somebody had to do it.' He put out a hand. 'Good luck.'

I was the only one with whom he'd shaken hands. The hangers-on from halfway down the crocodile looked at me with great respect. A couple of majors took my hand in passing. They probably thought I was a war hero, hurt in the battle for Furunculosis.

Change was everywhere when I got out of hospital with my carbuncles cured. Nobody seemed interested in punishing me for desertion. I was sent to a stores unit in Athens as mechanical transport officer. I was put in charge of over 100 vehicles. And I couldn't drive, and didn't learn until many years later.

My good right-hand man was a Maltese sergeant driver called Salvatore. Slim, wiry, he was all action. Driving had been his game throughout the whole war and he knew the business inside out. He'd joined up from Malta and had spent his time in the Middle East and Greece ever since. Back home he had a wife and an army of kids but that didn't stop Salvatore from rogering as many Greek women as he could manage. When I met up with him he was living out in a civilian house, against all regulations, sleeping with a mother and daughter. He wasn't bullshitting when he told me this. I visited with him at the house, and saw him fondle them both, carefully showing no more love for the one than the other. The daughter was a little dark-haired, fiery-tempered cracker. I tried to get organised with her behind Salvatore's back but she only looked at me with scorn. 'No man could be better than Salvatore,' she said, and that was that.

I hadn't been a month on the stores unit when the commanding officer, a genial character with a Mayfair manner and a fraudulent hyphen, called me in to see him. 'I don't know whether you know this, old man, but you came to me under a bit of a cloud,' he said, sitting back in his chair and studying me, not that his studying bothered me any. I was happy once again, running the lorry pool. The wild fidgets that had troubled me on my arrival in Greece seemed to have left. Indeed, I was quite the gentleman, on account of the fact that the unit stocked officers' drill uniforms, and I'd been fully kitted out, with all items tailored, following the simple expedient of letting the stores sergeant have a fifteen-hundredweight truck any time he wanted one. I must say I looked quite the part, that day in the spring of 1945.

'You've come through with flying colours, as far as I'm concerned,' the CO said.

'Good man yourself,' says I. 'And if anybody asks me about you I'll say the same.'

'Now,' the other man leaned forward, 'I have a problem which I believe you, with the same sharp brain that has talked the stores sergeant into kitting you out without cost, can solve.'

I studied him afresh. This was no mug.

'A letter being censored by one of our officers, written by an airman, turned out to contain an instruction to his family to get in touch with their local Member of Parliament.' The squadron leader sat back and waited for my reaction.

'Sure they're all at that nowadays,' I said, and so they were. All of a sudden the British armed forces were aware of their rights. With the war coming to a close, the authorities had instituted classes for the troops, at which current affairs and political history were debated, with other subjects. The idea was to prepare the men for release from the forces. There was a fear that they might go back and start riots if things didn't suit them. Another reason for the classes was to give the men some sort of outlet to compensate for the restlessness that inaction brings. The same sort of after-the-fighting jumpiness that had sent me off on a drinking tangent myself.

'This letter was complaining about the food,' the CO said, 'and it alleged that the officers were eating the men's rations.'

'And are we?' I asked.

'We're not,' I was told, 'and that's not a very funny remark. What's happening is that officers pay a little extra towards added rations, bought locally. Under King's Regulations the men aren't allowed to do this.'

My brain was working, but I could hardly believe what it was telling me. 'So you've brought me in here as just about the only man who can end this muttering in the cookhouse.'

The CO nodded.

'And, stop me when I'm getting too outrageous, you don't want to know too much about how the men's food is suddenly improved?'

'There have been several letters sent home complaining about the food on this unit. It's not my fault. You know as well as I do that the Germans took all the transport away with them when they went north. The farmers can't get their produce to the shops in Athens . . .'

'And I have over a hundred trucks,' I finished.

He shook his head emphatically. '*I* have over a hundred trucks, not you. But if you were to find a use for those trucks which does not lead to a court martial for yourself – it certainly won't for me, because I shall deny that this interview ever took place – then the cloud which you arrived under will turn out to have been summer cumulus.'

I sent for Sergeant Salvatore. 'Open that drawer of yours where I know you keep Scotch whisky,' I said. 'You and I have exciting things to discuss. If you were to have a wish granted, Salvatore, would that wish not be that you should go from here to some rural spot in the interior, fill up three trucks with produce, and bring it back here to help make the men's food more attractive, and to stop them whining?'

He sipped, thought for a minute. 'Forgive me, sair, but – I was listening extremely carefully just now – you didn't actually say whether it would take the whole three truckloads of stuff to stop them whining . . .'

'No, indeed, now you mention it, I didn't.'

Sergeant Salvatore touched my glass with his. 'Sair, I know what your ambition is. You have told me often enough. It is to get a woman who will be well brought-up, sweet and kind, and who loves it. If two buckshee truckloads of stuff don't get it for you then nothing will.'

I raised a correcting finger. 'One and half buckshee truckloads, Salvatore, not two. For every spud that we knock off and flog to the Athenian shopkeepers, a spud must go to my men. Throughout this war they have folded uniforms and packed socks in boxes. Do you think I would let them down now? It's spud for spud or nothing.'

Salvatore smiled. 'I was hoping only for half a truck,' he said. 'You have much to teach me, sair.'

I rose. 'Any man who can roll in the scratcher with two women at the same time is in no need of lessons from me,' I said. I dismissed him and finished his Scotch.

The very next morning we were on our way in the pilot three-ton truck. Down to the Peloponnese we went, past the shattered reminders of war, through villages where ropes which had hanged hostages still hung in the squares. The further we went from Athens, the more plentiful was the produce in the fields. We stopped, leisurely, ate grapes, drank retsina and chewed goats'-milk cheese in little tavernas. It was outside one of these, not twenty kilometres from Kalamata, that the two guerillas came upon us.

Greece, after indescribable suffering under German occupation, was torn apart by civil war, when the ELAS Communist guerillas tried to seize power. After bitter fighting, in which the British used Beaufighters to blast ELAS positions in Athens and the RAF headquarters buildings actually fell to the insurgents, an uneasy truce came about in January 1945, just before I arrived. But ELAS guerillas were still roaming the countryside. Obviously, Salvo and I had been found by two of them.

'Can you fire your Sten gun, Salvatore?'

He had spotted them. Still chewing the coarse bread and cheese, he lifted the Sten gun that all troops were required to carry outside barracks, slipped the catch off and stood up. I took my Smith and Wesson out of its holster. I had fired only a dozen shots from it, on a range in the desert. I'd been almost the worst shot on the squadron. One of the bullets, fired at twenty-five yards' range, had actually gone into the ground twelve yards in front of me.

The two men stopped, fifty yards off. One had a Lee Enfield rifle over his shoulder. Relieved, I saw that he intended to keep it there. The rifleman shouted something in Greek.

'They are saying to us to sit down and not to worry, sair. They say they just use their gun for robbery and not any more for revolution.'

'Well, what a bloody good idea.' I put my gun back where it belonged, waved cheerily at the Reds. 'Kalisperos,' I called, meaning 'good afternoon' in Belfast Greek.

Salvatore reluctantly sat down, but he kept his Sten at the ready. The two men came up to our table and stood before us smiling. They were in their mid-twenties, hard-looking merchants, plenty of growth on their faces, needing a bath badly. Behind us, the owner of the taverna had emerged. He stood watching, eyes narrowed against the sun.

The rifleman was the leader. He sat down at our table after nodding a question to me and getting my smile back in return. I called the proprietor and ordered drink for the two, with an ouzo for myself and a retsina for Salvatore. Then my sergeant began to talk to the strangers. The first bursts of Greek on both sides were interrogatory. Salvatore gestured towards our truck, lying ten yards away.

'I have told them that we are alone,' Salvatore told me, 'and that we do not want to do them any harm.'

'Great stuff,' I said. 'Keep it up.'

Salvo laughed suddenly and laid the Sten down at his feet.

'Tell me,' I said, 'I want to hear what's so funny.'

'They say that they saw from a great distance that we were RAF, and the RAF had been so obliging in the Rising, letting their headquarters be captured, that they knew they were in no danger.'

'Tell them why we're here, to buy goods and flog the half of them for our own benefit,' I said. 'Invite them in for a consultation fee if they'll put us onto the best markets.'

There was more talk. It broke off when both guerillas reached over and shook hands with me. I clapped for the taverna owner and ordered more drink and food. The talk was resumed.

'Exactly what sort of farmers do you want to meet?' Salvatore asked me.

'I want to meet farmers who have plenty of eggs and tomatoes and potatoes, and who are not very good at bargaining,' I said. When this was translated there were roars of laughter.

Salvatore had to stay sober to drive, but I was able to get roaring drunk with the enemies of the state. We swopped all sorts of recollections. They'd seen the Beaufighters over Greece during the occupation. The rifleman said that he had seen one attacked by an American Lockheed Lightning one day.

'Who won?' I enquired.

'The Britisher,' he said. There was a noisy cheer, and we all sang the 'Internationale' after a fashion. Then we went off with the guerillas to the edge of Kalamata, where they left us with hugs and handshakes. They'd given Salvatore the address of two women who would put us up and – wink, wink – look after us. Salvatore had told them of my quest for a nice girl with nympho inclinations. The women were nice all right, and I ended up in bed with one, but I was so full of ouzo that I flaked out until ten the next morning, by which time the lady had gone.

We came back, however, with the lorry full of eggs, tomatoes and potatoes. The supply line was laid down. The guerillas' rake-off was all organised, the pick-up point being the taverna where we'd met. All was set fair for the life of Riley, and it was just at that point that I was posted to Air HQ as the transport staff officer.

I must say that I arrived in Air Headquarters Greece a miffed man. Looking forward to the life of a Mafia boss, with a comfortable flat and an attentive, loving, savaging, biting girl beside me, I had been thrust behind a desk by cruel fate. Others were reaping the reward for my legwork. Still, once I settled into the staff job there were certain areas of satisfaction.

Firstly, Greece was riddled with rackets, run mainly by the British armed forces. Vehicles were obligingly left where the locals could knock them off; blankets were at a premium, not to keep the Greeks warm but to make nice clothes for the ladies. Parachute silk fetched high prices in the black market, as did mosquito netting, for curtains. The country was in an appalling mess. There was hardly any civilian transport. The Germans had even cleaned the place out of horses before leaving. I was virtually in charge of all the RAF transport in Greece. Admittedly each unit had its own transport officer, but I had a position which carried clout. It was my job to help stamp out the rackets, and there was hardly a transport unit of any service in Greece that wasn't saturated with villainy. I was ordered to go out, find it and fetch the cops. This meant that when I arrived on a truck compound a shiver ran through the place. The word was that the rogue airmen were being paid in gold sovereigns and, when the money was really big, that it was being lodged in Turkish banks, to be realised after the war.

'I don't suppose you'll know much about the black market,' the wing commander who was my boss remarked.

'Not a sausage,' I replied with a straight face.

'Go out into the world and help the Special Investigation Branch to save souls,' he directed.

Right enough, the SIB were breaking their necks trying to catch the chief operators. One way was to examine pay books and to closely question those who hadn't drawn their pay for months. Another giveaway was the number of savings stamps being purchased by certain of the troops. But one thing was made clear by me from the outset – I wasn't going to be any kind of Sherlock Holmes working for the Special Investigation Branch.

So what I did was to go along in a sudden swoop to a unit, examine the mileage returns and the consequent petrol usage, look piercingly into the eyes of the NCOs running the pools, and tell them that I was no mug and that whatever was going on would stop, or by God the military police would be in on the act. Then I would allow myself to be knocked blind drunk in the mess, and I would sing 'Danny Boy' upside down on the bar.

In this way I covered Greece from top to bottom and from east to west. I even realised an ambition and travelled over Mount Olympus by moonlight. I was in company with an air traffic control officer called Drewitt, who was going with me to Salonika, and Drewitt was a man who knew his Greek mythology. He was also a bit of a wit. 'Here we sit at the feet of the gods,' he said, 'listening to the sound of nanny goats' bells, and up to the ankles in nanny goats' balls.'

I was driven to write in a classical vein myself at this time. I had this character called McAuley, a Belfastman, who believed that he had had an earlier existence in which he was servant to the poet Appollonius Rhodius, the man who wrote the story of Jason and the Argonauts. McAuley believed that his name in this earlier life was Makarios. It began something like this:

With a screeching of brakes, the ambulance shuddered to a halt. Its doors flew open, and the victim, waxen-faced and still, was rushed into casualty. A young doctor peered intently into the eyes of the man on the stretcher:

'Can you hear me?'

One of the ambulance men touched the doctor's sleeve: 'It was a punch-up in the Gibraltar bar, down York Street.'

'Can you hear me?'

There was a barely perceptible nod. The doctor turned to the ambulanceman. 'There was a barely perceptible nod there, kid,' he said.

'Frigging good show,' was the reply.

'What's your name?' The doctor bent his head to catch the whispered reply.

A young nurse with large breasts came to watch. 'What's going on?' she asked.

'The doctor's bending his head to catch the whispered reply. You know, like this,' and the ambulanceman buried his mouth and nose in the nurse's bosom. She nodded. Taking the cigarette from her mouth, the nurse stubbed it out in the ambulanceman's ear. Just with that she noticed the doctor frown.

'I noticed you frowning there, doctor,' she said.

Looking up from the stretcher, the medico said, 'This guy reckons he's servant to the poet Appollonius Rhodius. He says his name's Makarios.'

'Christ, talk about being a manifest liar,' the ambulanceman said, cramming healing balm into his earhole.

'Why?' the doctor and the nurse asked together.

'I know him and all belonging to him,' the ambulanceman said. 'His name's McAuley and he's a boilermaker's helper in the shipyard.'

The young doctor was staring, hypnotised, at the nurse's bosom.

'So his name's breast, is it?' He shook himself. 'I mean, his name's McAuley, is it? You don't half meet some daffy down dillies in here, don't you?' He looked at his watch. 'Goodness, it's half-past breast already.'

He moved to the next patient, a circular-saw operator who, after losing two fingers a couple of months earlier, had just lost two more, showing a mate of his how it had happened . . .

11

Anna

Meanwhile I was racing through 1945 in great style, meeting all sorts of characters. Up in Salonika I was seconded for a spell to the Greek Air Force, which was just reforming as a peacetime force. I was living at Sedes airport outside the town. Through the day my job was to teach the pilots navigation. The first exercise was planned to take their Spitfires close to the border with Bulgaria. Off they went, and I must admit I wondered at their wild enthusiasm. They were laughing and jumping about like schoolboys on the way to the flights.

When they got back I found out why. The whole formation had departed from the flight plan as soon as they'd reached the border. Over the line they'd gone, down on the deck, and they'd played merry hell beating up towns and villages, sending terrified Bulgarians scampering for cover in all directions. Protests were pouring in for weeks afterwards.

That's when I first learned that the Greeks hated the Bulgarians with a deep and lasting vigour. Later I learned also that the Greeks hated the Albanians every bit as much. And the Turks. Boy did they hate the Turks. Every evening by the waterfront in Salonika the men would walk up and down, up and down, in twos, thumping fists into palms of hands, shouting, arguing, nearly crying with rage, because of what the Turks had done to Greece in an earlier occupation of the country.

Then, of course, there was the question of Cyprus. It was beginning to lift its head in 1945, although no shots were fired then. That was to come later. But *enosis,* or union with Greece, was another word that fairly rang around Salonika harbour of a spring evening, when I was teaching the Greeks to navigate.

But I loved the Greek people. In many ways they reminded me of the Irish. They were great fighters, none better, and they had suffered because of other nations' greed. I first took to the Greek people within an hour of landing in the country from Egypt. I was walking through Athens with a tall cockney warrant officer pilot when we were approached by a young local man, selling something or other from a tray hung round his neck.

In Egypt one had to be very careful of such guys. They tended quite often to be pickpockets. The speed of the hand deceived the eye. One quick flick of a dirty postcard before the face and the wallet was among the missing. Some of the troops were inclined to be less than gentle with street salesmen. The cockney pilot with me was one of the physical kind. When the Greek persisted with his sales patter my companion, as to the imperial manner born, simply took the back of his hand and pushed the other in the chest. His victim stumbled, fell against the wall, and tipped the contents of his tray all over the pavement.

Although the other man tried to prevent me, I bent to help pick up the few cheap combs, brooches and other items, but the Greek, as soon as he got his breath back, took off like a greyhound from the trap and, before the cockney realised what had happened, he'd copped a beauty right up the mouth. It took me and a couple of passing servicemen to pull the two apart.

'By God you're not in Egypt now, son,' I chortled, once we got under way again. 'You made the same mistake as Mussolini did with these guys, and you got the same medicine.'

Many times, in different parts of the country, I noticed the same fierce individual pride demonstrated by the Greek people. Their touchiness used to irritate many of my comrades, but I loved it, having been born and brought up myself in a community where skins were thinner than the water round a soap bubble.

In the bombing of Sedes airfield, Salonika, the Allies had, *inter alia*, flattened a brick-built hangar. When I got to Sedes in mid-1945 it had been in Allied hands for seven months, so there was no reason to believe that essential tidying-up, like the burial of dead bodies, still remained to be done. But I found out differently when, passing the huge mound of bricks and rubble that had once been the hangar, I idly skimmed a stone into it.

The stone landed with a clatter, and I stood nailed to the spot in horror. The whole face of the mound seemed to move and writhe as literally hundreds of rats scurried and scampered in and out through the broken masonry. At the same time I heard the most revolting scrabbling and scraping of claws, together with a chorus of squeaks and cries that sounded like something out of a Dracula film.

'Sweet Christ,' I said aloud. There must have been one hell of an attraction at the bottom of the heap to make such an army of rats so angry at being disturbed.

A second later I nearly lost my reason as, from somewhere behind me, someone screamed.

As usual when under severe stress, I broke out in severe swearing: 'Holy fuck!' I yelled, and swung round to fight the vampire bats.

To my relief all that I saw was a young girl, standing a dozen yards behind me on the cracked tarmac. She was biting her knuckles, staring in hypnotised terror at the seething, boiling rat-face of the bombed hangar. Then she ran towards me. I put out my arms and held her, as the squeaking and rustling lessened. 'Holy fuck,' I said, tucking her in nice and tight, stroking her hair, giving the nearest of her ears a bit of a nibble, just to remind her of the terror nearby and to keep the fright as an external force pushing her pelvis smartly into mine. I was particularly delighted to notice that no one else was near. The spot was well away from the centre of airport activities, and I was only there in the course of my usual hoking and poking around battle scenes.

The girl drew away. She pointed to the rats. 'Holy fuck?' she said, in a strong Greek accent. She obviously thought that this was the English for rats.

I nodded. There would be time for the jolly old language lessons later, after I'd milked her fear to the uttermost midriff shiver. I looked closely into her frightened face, and at once all the lust died. 'Well, holy fuck,' I said, in disgust.

She was only a schoolgirl. About fourteen at the outside. Her eyes were round and innocent. The fear that showed there was a child's terror at things that belong in the dark.

I pushed her away, gently. Then I made to walk on, but the Greek girl clung to my arm as we skirted the probable burial-place of a number of Luftwaffe flight mechanics. It was a good half-mile to the control tower. I made talk as best I could on the way, saying things like, 'You were near yourself there, for a minute, wee girl,' and, 'Have ye any big sisters back home?' Reaching the tower, I shook hands in the Mediterranean way, with liquid eyes and a melting smile, then I left her, and the effect on my sex-pack of our meeting wore off in about two hours.

That very evening, I fell in with a drunken armament officer name of Martin. 'Come with me,' said Martin, 'for I know whereof there is free brandy to be had.'

'You've talked me into it,' I replied, and we set off.

He brought me to a roadside taverna. Martin fetched the proprietor to me. 'This is Demo,' he said, 'and Demo has a most interesting tale to relate. I have told him that you are the official war historian for the entire Royal Air Force in the Central Mediterranean theatre and a confidante of Field Marshal Alexander.'

'Good wee while since I saw Alex,' I said, shaking Demo's hand. I pulled a pen out, and a letter from home. 'Dear Sam,' the letter began, 'just a few lines hoping to find you well, as this leaves us all well at home. Well, son, and what do you think, but Jim has bought a greyhound. It lost its first race, but he says we should all back it the next time out. It only lost, he said, because it ate the straw in its bed before Jim could stop it.'

'From Air Marshal Sir Keith Park,' I explained. Then, poised to take notes, I looked enquiringly at the Greek.

'I have keep zees Martell brandy for ze day ze Hellenes have zere own Air Force and zere own Army,' he said.

Martin and I looked at each other excitedly. 'Why,' we said together, 'that's this very day. What a coincidence.'

The Greek was beside himself with joy. 'Ze Bulgars!' He held a hand out, finger extended.

'Boom-boom-the-Bulgars,' I shouted.

'Up their flue,' Martin yelled.

Demo broke out the Martell brandy. We sank it as if it was best bitter, Martin and I. I began to build up a friendship with the Greek. Everyone in the country seemed to have an interest in merchant shipping, and I explained carefully that my father owned a small line called the Head Line. Only sixteen vessels, I said, but growing.

We were talking about the effect that the war might have on seaborne commerce, when through the door walked the girl I'd met by Ratsville a few hours earlier.

'Well, shiver my timbers, rinse me and rot me, if it isn't my little friend,' I said, rising.

She came forward eagerly. I drew up a chair, and started to tell Martin and the Greek about our experience, when I noticed the girl and Demo exchanging surprised smiles. He put his arm about her shoulders. 'Anna is mine child,' he said proudly.

'Not child. Girl,' said Anna.

'She told you about today?' I was a little worried. Girls weren't daft. She must have reflected on the amount of voltage she'd generated in me on the airfield, never mind that she was just fourteen. But there was only fondness in her eyes.

'You stay with me for dinner.' Demo was adamant. I agreed.

Martin, as was his usual style, flaked out suddenly. He was carefully laid out on a long couch, and the three of us moved into Demo's comfortable house next door. He and Anna comprised the family. Demo told me on that first evening that his wife had died early in the war, of blood poisoning, following a bad gash from barbed wire, when she fell into a ditch by the airfield perimeter. He had been the supplier of local wine to the men's canteen on Sedes all through the occupation, and his wife had been returning in darkness after taking a weekly order when the accident happened. No doubt it was the effect of malnutrition that finished her. Demo and his wife had been careful to live on the same bare allowance of food as the rest of the local people, since it would have been a hanging matter if they had accepted German gifts of food. Anna, his daughter, had been looking for the men's canteen that afternoon, when we'd met.

During the evening, I couldn't stop thinking about the girl, although I kept my eyes everywhere but on her. She was small and had eyes that were golden light-brown. Most disturbing eyes they were. Her figure was gorgeous. In every regard Anna was a woman, except for her age. Her father explained to me that she did not go to school because of the bad times, but that she certainly would be going as soon as the country was back on its feet. He wanted Anna to be a schoolteacher. 'Do you think she will be nice schoolteacher?' he asked.

I know what she'd be nice at, I thought to myself, and it's a long way from algebra. More like multiplication. I shook hands with Anna solemnly, was hugged by Demo, and left, pissed as the duck that drank out of the hops bucket.

When I woke up next morning I could hardly live with myself. I considered myself more than halfway along the road to perversion. It was criminal to take unto oneself a fourteen-year-old girl. It was bound to be dangerous even to think about it. And in Greece it was next door to suicidal to let the gaze linger for more than one thousandth of a humming bird's flutter on a female child with her old man looking on. It didn't take a mighty effort of self-control to resolve never to darken the door of Demo again, Martell or no Martell. I had arranged to call at the taverna. I found no difficulty in breaking the promise.

A week went by. Once again Anna's effect on my near-virgin system had abated, and I was at work in my office in the instructional block on the airport. I was plotting a course for the Greek Spitfire pilots that would lessen the chances of an attack on Bulgarian sovereignty, when there was a knock on my door. An airman opened it, peered in, leered, jerked his head back over his shoulder, and then Anna walked in.

She was shy. She was nervous of her temerity in coming to see me. She was unsure of me. All of a sudden I was the British officer, sober, correct, surrounded by maps and charts, someone whose door was tapped respectfully by an airman. All these things I could see in her face. But there was something else I could read there. As I came round from behind the desk and stood near Anna, I could see tiny sweat-beads on her upper lip. There was something in her expression, a sort of hunger. I took her hands. They were clammy. She was trembling slightly. I walked over and locked the door, then I came back and took her in my arms, and, together, after the first kiss, we sighed all the way down to our toes.

She couldn't kiss. She knew nothing about lovemaking, but even so I was nearly demented for her. We kissed again and again and stood close. Then I led Anna out, along the corridor, through the door, and all the way to the main gate. Demo's taverna was only yards away. Together we walked there. As we neared it, Anna drew her hand away from mine. Her expression as she did was conspiratorial.

I stopped Demo from rising. 'I have come to say cheerio, Demo,' I said, 'and I cannot say the words to Anna. Will you say it for me?'

There was a long pause. Demo looked closely at Anna.

'Say goodbye for me,' I said once again.

He said something in Greek. Her head jerked to me in astonishment.

I took Anna's hand. 'Goodbye, Anna. I go to Athena.' I left her then.

She stood awkwardly, like the schoolgirl she was. Her face was crumpled. Her mouth twisted sulkily.

Demo saw me to the door. 'You are young *officieri,*' he said. 'She like.'

I shook hands and walked away.

Says I to myself, she doesn't like even a tenth as much as I like. Just my frigging luck. Just another two years would have done it . . . That afternoon I asked, urgently, to be sent back to Athens.

One day I happened to be in the centre of Athens, near the Grande Bretagne hotel, which was the main billet for British officers, when I saw one of the vehicles from the stores unit where I'd once worked being driven by a Greek Air Force sergeant. I was with an Irish navigator from County Antrim at the time. 'Wait a wee minute,' I said, and signalled the truck to stop.

'Why you drive this truck?' I asked, leaning into the cab.

'Why you ask?' was the reply.

'Me big transport *officieri,*' I replied.

'*Officieri* fick off,' the Greek said, letting the clutch in.

I jumped inside and tore the keys from the ignition. 'You speak me,' I shouted. 'Show me permit.'

The Greek nodded resignedly. As he climbed meekly down onto the pavement, I couldn't help shooting a mildly triumphant glance towards the County Antrim man. He was passing through Athens on his way home from the Far East, where he'd been flying Beaufighter torpedo bombers. It was satisfying to let him see that I was some administrative whizbang, as well as a flier. He looked suitably impressed.

'Show me permit, please,' I said to the Greek sergeant.

'Here permit,' he said, producing a starting handle from behind his back and sending it like whispering death in the direction of my skull.

I ducked, and the pair of us got stuck into it there on the most fashionable street in Athens. In no time at all a huge crowd had gathered, the Greeks cheering their man on and the troops yelling their lungs out for me.

Two huge American GIs began to lay bets on the outcome. I managed to wrest the starting handle from the other man. 'Throw it away and fight clean, Sam,' the County Antrim man shouted.

'Away and frig yourself,' I said, and let the Greek have the angle of it round the ribs. 'I got ten bucks here says the Limey kills him,' a Yank screamed.

'Take evens to that ten bucks,' I instructed my compatriot, shifting the starting handle to the other hand and whaling the sergeant in or about the left shoulder. That did him. He threw up the hands and the fight was over. After being asked for the third time to produce a permit for driving the truck, he reached into a pocket and did just that. Then he went back and complained to his superiors.

'It says here,' the wing commander said the next day, 'that you attacked this sergeant with a starting handle for no reason at all.'

'Jaze,' says I, 'that's a good one.' And I gave him the full details.

'Have you no pride in the Royal Air Force?' he asked.

'It depends on the outcome of this enquiry whether I've any pride in the Royal Air Force or not,' I said.

The wing commander shook his head in bewilderment. 'I don't honestly know what to do about you,' he said. 'I might as well tell you I've had the most amazing stories passed on to me about you, stories from all over Greece. Did you have some sort of a scuffle with a German officer prisoner at Corinth?'

'You wouldn't exactly call it a scuffle,' I said. 'What I did was to knock him on the flat of his back. Then he knocked me clean out.'

The senior officer's mouth fell open. 'Great God in heaven,' he said, 'I can't bring myself to believe this. A British and a German officer fighting like private soldiers.'

'Well, I was in the queue for dinner, in this second-line Army workshop near Corinth,' I said, 'and this German jumped in in front of me. "Back of the queue, Adolf," I said, and pushed him out. He pushed me back. I asked him who exactly he thought was winning the frigging war, and handed him a fourpenny one across the side of the gub. Next thing I knew the men behind me in the dinner line were burning paper under my nose to bring me round. Apparently,' says I, 'over on the islands where this German spent the war they did nothing but run boxing tournaments, but how was I to know that?'

'And what action did you take in regard to this prisoner who struck an officer?'

'What action do you think I took. I let him into the queue in front of me.'

The wing commander stared thoughtfully at the file before him. 'Did you know that someone back in the UK is making representations to have you sent home?'

'No, I didn't.'

It had to be Uncle Alec.

'Well, I might as well tell you that I'll be doing what I can to support any efforts to send you out of Greece. But if you're still here in a month's time, when I'm required to give a probation report on you, you'd better look out. Now clear off.'

A good friend of mine was in the military hospital in Athens with glandular fever. When I left the interview I called to see the patient. We were chatting about one thing and another, when the door to his room suddenly burst open and an American flier charged in.

'Say, you guys, guess what?'

My friend was a product of Charterhouse and Cambridge. He had the sort of fine hair, falling carelessly over the brow, that British royalty had. His eyes were blue and his skin was almost creamy. If he hadn't been a thoroughly entertaining character I'd have liked him anyway, for his upper-class mannerisms, quite naturally displayed. He broke off our conversation and turned his glance to the Yank: 'Yes, old chap?'

The visitor was almost beside himself with excitement. 'Yippee!' He jabbed a fist in the air. 'What the hell d'ya think, fellas?'

My friend's breeding prevailed. I kept my curiosity in check. Neither of us spoke. The American was purple with the tension.

'The goddam war's over. The Heinies have surrendered. It's on the radio. The war's over, guys!' Panting, he waited.

'Really. Jolly good of you to let us know, old chap.' And the Englishman turned back to me. 'What were you saying, Paddy?'

The American stared unbelievingly. Then he turned and went out, shaking his head.

And that's how I learned that our war was over, and that I was going to make it home. Outside in Athens they were dancing in the streets, soldiers were kissing girls, bartenders were giving away free drink. My friend and I talked about times past, on the squadron, in England, in Canada. I'd been his senior cadet at St John's Wood when he'd first put on a uniform, and we'd liked each other's company ever since. When we parted on VE Day he gave me his home address and made me promise to look him up after we got back.

Of course I never did. It would have turned out something like *The Admirable Crichton*. The butler might be king on the desert island, but back home he's still the butler. I would have nothing in common with stockbrokers, where I was going. Still, I was glad that I'd been with an old comrade in arms when I'd got the news.

I heard no more about Uncle Alec's manoeuvres on my behalf for some months, and meanwhile I carried on as before, with no follow-up to the wing commander's threat. During the second half of 1945 I stayed mainly in and around Athens. With the constant shortage of transport, visiting officers were ringing up from the airfield on arrival in Greece, asking for cars to be sent for them. Sooner or later their requests reached me, wherever I was.

'Hello, is that the transport staff officer?'

'The very one.'

'Well this is Squadron Leader Hodge. I'm the National Savings Officer for the Central Mediterranean. I want collected here at the airfield and brought to my hotel.'

'Sorry, sir. No cars available. Next bus for the troops is in three hours.'

'Look here, what is your rank?'

'Flying officer, sir.'

'Well, you're addressing a senior officer. I have no intention of travelling in a bus meant for the troops. Send a car for me at once, do you hear?'

'Haven't got one.'

'I'm ordering you.'

'Order away there.'

'Are you disobeying an order?'

'Yes. Isn't it desperate? Bye-bye.'

I had all sorts of satisfying moments like that. But there was the postwar feeling of frustration eating into everything. I didn't even get any satisfaction out of inspecting the guard any more. The fun of selecting the stick man had almost evaporated. This was an old Army tradition, taken up by many RAF units. There was always one extra man on the evening guard parade, and the orderly officer, after his inspection, awarded the stick to the best turned-out man. This meant that he was excused guard for the night – a considerable honour.

Usually the stick went to robots, who'd spent hours polishing and buffing and ironing. I awarded my stick to the scruffiest bastard on parade. The sergeants and warrant officers went daft, but I was adamant. As soon as word got out, my guard parades began to look like the end-of-the-pier show. It was marvellous. I was actually shaping the Air Force according to my wishes. My heart used to sing, watching my bunch of ragtag and bobtail shuffle past the flagpole as the RAF standard was lowered at the end of the day. Sometimes, during an evening on duty, I would call into the sergeants' mess and start an argument, as in days of old. Once or twice I even got the hands up. Blows were struck. Lovely stuff. But it was all dying down now. *Esprit de corps* was escaping, hissing out of the corps like air from a punctured tyre.

The signals from Uncle Alec's direction back in Belfast began to grow stronger. He was getting closer, like someone outside prison, tunnelling in under the wall. 1945 came to its end. Greece looked as though she was in for another rough time of it. Rumours of another rising were rife. Sergeant Salvatore, who was walking around like Al Capone with gold rings and bracelets everywhere, told me that the food runs had had to be stopped. The guerillas weren't nearly so friendly and, more to the sinister point, they had stopped robbing for robbery's sake, and were shovelling money into the cause. Everything was changing. Certainly the money was changing. Every few weeks the bartenders would add a nought to the number of drachmas needed to buy drink.

My hopes for a nice woman who would be good, intelligent company and think of nothing but sex had been cruelly dashed. Dire warnings by friends in the medical profession of VD amongst dance-hall girls had thrown me into civilised company. Through my links with the Greek Air Force I was being invited to respectable homes, meeting well-brought-up girls.

There was no doubt about it, there were some crackers there. And one in six could rub along with me. The trouble was, they went in for chaperoning in the fiercest way. And their big brothers? Jaze, I had a sister myself back in Belfast, but if I'd stuck my nose into her business the way the Greek brothers did, she'd have belted me one with the rolling pin.

'My leetle seester Helena, she likes you vairy mooch, Paddee. You like her mooch?'

'Well, all right, you know, Skenas. OK, like.'

'You like get married, Paddee?'

'No. No like get married no time. Never.'

'Zen you not touch my leetle seester, Paddee. OK?'

Bloody marvellous. Roll on the boat. Let's launch this pent-up passion where it never should have left – the waterfront of Belfast, Ireland.

Early in January 1946 I was sent for by the camp commandant, a man who had flown many operations with me.

'There's been action on the compassionate leave front, Paddy.'

'What way, action?'

'Things have moved.'

'Am I going to move?'

'If you so wish.'

'Compassionate leave? I thought there was no such thing.' I recalled some of the most heartbreaking cases among the ground staff in the desert. They'd all had their requests turned down. One man, a fitter, produced for my benefit one day a letter from his mother-in-law telling him that his wife was so far gone with cancer that she weighed only five stones. The fitter had showed me a snapshot of a fine, big, laughing girl on the beach, wearing a Kiss Me hat. I'd fought his case but nobody ever got compassionate leave, I was told. The fitter didn't see his wife again.

'It's different for officers,' I was told by the camp commandant. I'd already found that out, in William Price's case, of course, and shouldn't have been surprised by the extent of the difference.

My mother was ill. They were going to send me home. A seat was available on a plane leaving for Pomigliano airfield in southern Italy. It seemed that the entire resources of Air Headquarters Greece were thrown into the business of getting shot of me. I didn't know for the next few hours if I was on my head or my heels, whether I was eating my dinner or the *Irish News*. I was hustled here there and everywhere, getting my clearance chit signed, handing in equipment, saying goodbye to all those I could manage to see.

'Am I not even going to throw a farewell party?'

'Every bloody party you've thrown for the last twelve months has been a farewell party. The ones drinking with you said never again. Hurry up. You've only time to pack and run.'

I ended up navigating the old communications Avro Anson to Bari and Pomigliano. It was the last time I would ever plot a course. There was no need to sweat over the airmanship in peacetime flying conditions, but I did it impeccably anyway, arriving over base on the button, running my hand fondly along the trailing edge of the port aileron as I walked away from the Anson.

I spent one night in transit accommodation at Pomigliano and got onto a DC 3 next morning for Blackbush, outside London. To my great surprise I was cleared of formalities in time to catch the Stranraer boat-train from Euston. Passing through Westminster I drew a fistful of pound notes from Cox's and King's bank. That roll of notes was the first of the big differences in the way I joined the 1939-45 war and the way I came home from it.

The Larne boat was still the same. We were still bunged in like cattle. Only on this trip I wasn't cattle. I went up to a steward. 'What I'd like's a room somewhere, where I can't be disturbed. Where I can a drink in peace and quiet.'

'You'll be lucky.' He sounded like a Newry man, spitting the words out like bullets from a Smith and Wesson.

I produced the roll. 'Tell me this, do you like one-pound notes, or five-pound notes?'

He reached out two fingers. I placed a fiver in the cleft that they made. 'Come with me, sir.' I was led to a comfortable room with a bunk bed and a table and chair. 'Bring me six bottles of stout, two separate glasses of whiskey, and take a bottle and a half 'un for yourself.'

'Certainly, sir.'

I lay down on the bed, kicked my shoes off, put my hands behind my head.

Mark Kerr died in the North Sea in the middle of 1944. When he got back from Egypt he'd written to me: 'The Derry Air Training Corps have asked me to give them a talk. I feel such a fraud, having done nothing so far . . .'

When he went to join his squadron on the east coast of England, he and Len, his pilot, were put straight onto operations. It couldn't have been more than his third or fourth when the flak, or fighter or whatever, got them and they went in. I supposed he'd been airsick to the last.

Mick McDermott was lost in one of those stinkers around the Adriatic. I'd found out something about Air Ministry procedure in such cases when I'd written from the desert to Mick. My letter came back marked, 'Return to Sender on Air Ministry Instructions'. I remember thinking, 'So that's how they go about it, eh?'

It had given me a curious feeling, over and above the loss of Mick. It had made me feel as though I'd inadvertently opened my personal file or something. This 'Return to Sender' business was something it was as well for us not to know. But anyway, Mick was dead.

Jack Burgess ditched once in the North Sea and got out with only a couple of burns to the face. Then he was sent home on leave, wrote to me crowing about his superior operational experience, went back and was killed by a night-fighter over Essen. His pilot radioed that he was bringing back one dead crew member. Then the plane spun in over its own airfield and all aboard were lost. He was killed in April 1943. His mother wrote to me from her house in west Belfast several times afterwards. I was a link with Jack. She used to say in her letters: 'Do you remember the way you and Jackie used to . . . ?'

If I remembered any one thing more than another about Burgess it was that he had always, always, got the better-looking woman.

Alec McCleod was the first of the whole lot to go. When he was training as an observer he used to fly over his beloved Skye, in Wellington bombers. He told his brother that it always made him feel homesick. I knew about that feeling. I remembered, far out in the Atlantic, turning for Islay, calling for a direction-finding station located in the Castlereagh Hills to the east side of Belfast, and hearing the operator reply. All the way across 500 miles that signal came, and I swear I smelt soda bread baking as the Morse reached my earphones.

They put Alec onto Stirling bombers. They were notoriously poor in a climb. Too much armour. Leonard Cheshire, the VC, made an aeroplane out of the Stirling. He ripped the armour plating out of it. But that came too late for Alec McCleod. One night, on a long run to bomb Naples, his aircraft, fully loaded, flew into a mountain in France.

Going out to find the ship's toilet I came across a flight sergeant observer, pushed up against the wall, looking cheesed off in the packed saloon. I brought him into my room and gave him half my drink. We talked for the rest of the voyage about nothing but women. The observer had flown in Lancasters all over the Reich. He could hardly hold his cup of Guinness for the shakes.

My brother met me at the railway station on York Street, the same one from which I'd left to go to war six years before. 'What are you going to do now?' he asked.

I told him that I had £400 in the bank. 'That's fifteen pounds a week for six months,' I said. 'So come back to me in the month of July, when the £400 has been blitzed in McGrane's pub. I'll tell you then what I'm going to do.'

The second day at home I was sitting in the Gibraltar bar on York Street, lingering over a pint, when a man who'd missed the war gave me a blank form. 'Fill that in,' he said. I looked at it. It was for the B Special Constabulary. 'That's in case there's a real war,' he said.

I was back in Ireland all right.

'Do you know, I don't think I'll bother,' I said. 'For the next six months I'm going to be busy drinking. But I'm keeping five pounds for dancing lessons too.'

'What do you want to take dancing lessons for?' he said.

'To get a woman,' I told him. 'I've been working it out and that's the quickest way. As I've been finding out for the last six years, putting on a uniform certainly isn't.'